HANDBOOK OF SEPARATION PROCESS TECHNOLOGY

Edited by
Levitah Castil Mapatac

Delve Publishing

Handbook of Separation Process Technology

Edited by Levitah Castil Mapatac

Edition 2018

Delve Publishing

Published by
Delve Publishing LLC
708 3rd Avenue, 6th Floor
New York
NY 10801
United States of America
www.delvepublishing.com

ISBN: 978-1-68095-730-3 (Hardback)

Printed in United States of America

Library of Congress Control Number: 2016920410

About the Editor

Levitah Castil Mapatac

Levitah C. Mapatac is an Associate Professor, researcher, writer and a PhD holder from the University in Science and Technology of Southern Philippines. Her interest varies from different chemistry concepts and applications and currently doing research in traditional medicines, water waste pollution and environmental chemistry in the aspect of contamination pathways, profilling and rehabilatation of mining waste.

Table of Contents

3 CENTRIFUGATION AND CYCLONIC SEPARATION

4 CRYSTALLIZATION PROCESS

5 DRYING TECHNIQUES

6 EXTRACTION CHEMISTRY

7 THE TECHNIQUE OF FROTH FLOTATION

8 NATURAL-GAS PROCESSING

List of Abbreviations

AMD	Acidic Mine Drainage
CAE	Capillary Array Electrophoresis
CSP	Chiral Stationary Phase
CIP	Cleaning-in-place
CCC	Countercurrent Chromatography
DGF	Dissolved Gas Flotation
DL	Double Layer
ESI	Electrospray Ionization
EBA	Expanded-Bed Adsorption
FPLC	Fast protein Liquid Chromatography
FC	Forced Circulation
GPC	Gel Permeation Chromatography
HIC	Hydrophobic Interaction Chromatography
LLE	Liquid–Liquid Extraction
MIBC	Methyl Isobutyl Carbinol
MEPS	Microextraction By Packed Sorbent
PAGE	Polyacrylamide Gel Electrophoresis
PEG	Polyethylene Glycol
PAX	Potassium Amyl Xanthate
PIBX	Potassium Isobutyl Xanthate
PTV	Programmable Temperature Vaporizer
RPC	Reversed-Phase Chromatography
SMB	Simulated Moving Bed

SEC	Size-Exclusion Chromatography
SEX	Sodium Ethyl Xanthate
SIBX	Sodium Isobutyl Xanthate
SIPX	Sodium Isopropyl Xanthate
SPE	Solid-Phase Extraction
SPME	Solid-Phase Microextraction
SERS	Surface Enhanced Raman Spectroscopy
TBP	Tributyl Phosphate
TMBC	True Moving Bed Chromatography
UDDT	Urine-Diverting Dry Toilets

Preface

A separation process is a method to achieve any phenomenon that converts a mixture of chemical substance into two or more distinct product mixtures, which may be referred to as mixture, at least one of which is enriched in one or more of the mixture's constituents. In some cases, a separation may fully divide the mixture into its pure constituents. Separations differ in chemical properties or physical properties such as size, shape, mass, density, or chemical affinity, between the constituents of a mixture. They are often classified according to the particular differences they use to achieve separation. Usually there is only physical movement and no substantial chemical modification. If no single difference can be used to accomplish a desired separation, multiple operations will often be performed in combination to achieve the desired end. Separation processes are essential to the chemical, petroleum refining, and materials processing industries. The word "separation," however, refers to different processes and functions for different industries. Separation processes comprise a large portion of the activity in the chemical and petrochemical industries. In the forest products industry, separation ranges from separating wood chips of different sizes to converting black liquor for reuse. In the aluminum and steel industries, as well as the metal casting and glass industries, the most important area of separation technology is separation of different types of scrap, sand, and cullet. Finding common ground among these definitions of separation processes or technologies is a difficult task. Separation processes can be thought of in general terms as performing similar functions in all industries. This Text presents a comprehensive introduction of historically important separation processes such as distillation, absorption, extraction, leaching, and crystallization and considers these techniques in light of recent developments affecting them.

Content Coverage

Chapter One introduces the importance of separations, the concept of equilibrium, mass transfer, and problem-solving methods.

Chapter Two focuses on the process of adsorption that creates a film of the adsorbate on the surface of the adsorbent. This process differs from absorption, in which a fluid (the absorbate) is dissolved by or permeates a liquid or solid (the absorbent), respectively.

Chapter Three outlines the centrifugation and cyclonic separation processes. Centrifugation process is used to separate two miscible substances, but also to analyze the hydrodynamic properties of macromolecules.

Chapter Four highlights the crystallization process that refers to the formation of solid crystals from a homogeneous solution. It is essentially a solid-liquid separation technique and a very important one at that. Some of the ways which crystals form are through precipitating from a solution, melting or more rarely.

Chapter Five focuses on the mechanism of drying consisting of the removal of water or another solvent by evaporation from a solid, semi-solid or liquid.

Chapter Six emphasizes on extraction process consisting in the separation of a substance from a matrix. It includes liquid-liquid extraction, and solid phase extraction.

Chapter Seven focuses on the process of froth flotation for selectively separating hydrophobic materials from hydrophilic. This is used in mineral processing, paper recycling and waste-water treatment industries.

Chapter Eight presents the study of natural-gas processing designed to clean raw natural gas by separating impurities and various non-methane hydrocarbons and fluids to produce what is known as pipeline quality dry natural gas.

CHAPTER 1

Separation Process: An Introduction

Chapter Contents

- **Inroduction**
- **Separation Techniques**

Since separations are ubiquitous in chemical plants and petroleum refineries, chemical engineers must be familiar with a variety of separation methods. This chapter introduces the importance of separations, the concept of equilibrium, mass transfer, and problem-solving methods.

INRODUCTION

A **separation process** is a method to achieve any phenomenon that converts a mixture of chemical substance into two or more distinct product mixtures, which may be referred to as mixture. at least one of which is enriched in one or more of the mixture's constituents. In some cases, a separation may fully divide the mixture into its pure constituents. Separations differ in chemical properties or physical properties such as size, shape, mass, density, or chemical affinity, between the constituents of a mixture. They are often classified according to the particular differences they use to achieve separation. Usually there is only physical movement and no substantial chemical modification. If no single difference can be used to accomplish a desired separation, multiple operations will often be performed in combination to achieve the desired end.

With a few exceptions, elements or compounds are naturally found in an impure state. Often these impure raw materials must be separated into their purified components before they can be put to productive use, making separation techniques essential for the modern industrial economy. In some cases, these separations require total purification, as in the electrolysis refining of bauxite ore for aluminum metal, but a good example of an incomplete separation technique is oil refining. Crude oil occurs naturally as a mixture of various hydrocarbons and impurities. The refining process splits this mixture into other, more valuable mixtures such as natural gas, gasoline and chemical feedstocks, none of which are pure substances, but each of which must be separated from the raw crude. In both of these cases, a series of separations is necessary to obtain the desired end products. In the case of oil refining, crude is subjected to a long series of individual distillation steps, each of which produces a different product or intermediate.

Separators are used for the separation of liquid. In fact, they are vertically supported centrifuges and are built with flying bearings. They work with centrifugal force. A separator belongs to the continues sedimentation centrifuges. Both the purified liquid and the liquid separated from each other are continuously discharged by using a pump (under pressure) or pressure free. The solid material can be discharged discontinuously (chamber drum, solid walled disc drum), pseudo continuously (self-cleaning disc drum) or continuously (nozzle drum). The drum is the centerpiece of the separator, in which the separation process takes place. There are two types of drums: the chamber drum (known as chamber separators) and the disc drum (known as disc separators). The power transmission on the spindle and thereby on the drum can take place by using one of the three drive motors: helical gears, a belt drive or direct drive, via a special motor. The sealing of the separators is differentiated into four types: open, semi-closed, hydro-hermetic (sealing of the product space) or fully hermetic (absolute airtight).

The purpose of a separation may be *analytical*,can be used as a lie components in the original mixture without any attempt to save the fractions, or may be *preparative*, i.e. to "prepare" fractions or samples of the components that can be saved. The separation can be done on a small scale, effectively a laboratory scale for analytical or preparative purposes, or on a large scale, effectively an industrial scale for preparative purposes, or on some intermediate scale.

LIST OF SEPARATION TECHNIQUES

- Adsorption, adhesion of atoms, ions or molecules of gas, liquid, or dissolved solids to a surface
- Capillary electrophoresis
- Centrifugation and cyclonic separation, separates based on density differences
- Chromatography separates dissolved substances by different interaction with (i.e., travel through) a material
- Crystallization
- Decantation
- Demister (vapor), removes liquid droplets from gas streams

- Distillation, used for mixtures of liquids with different boiling points
- Drying, removes liquid from a solid by vaporisation
- Electrophoresis, separates organic molecules based on their different interaction with a gel under an electric potential (i.e., different travel)
- Electrostatic separation, works on the principle of corona discharge, where two plates are placed close together and high voltage is applied. This high voltage is used to separate the ionized particles.
- Elutriation
- Evaporation
- Extraction
    - Leaching
    - Liquid-liquid extraction
    - Solid phase extraction
- Field flow fractionation
- Flotation
    - Dissolved air flotation, removes suspended solids non-selectively from slurry by bubbles that are generated by air coming out of solution
    - Froth flotation, recovers valuable, hydrophobic solids by attachment to air bubbles generated by mechanical agitation of an air-slurry mixture, which float, and are recovered
    - Deinking, separating hydrophobic ink particles from hydrophilic paper pulp in paper recycling
- Flocculation, separates a solid from a liquid in a colloid, by use of a flocculant, which promotes the solid clumping into flocs
- Filtration – Mesh, bag and paper filters are used to remove large particulates suspended in fluids (e.g., fly ash) while membrane processes including microfiltration, ultrafiltration, nanofiltration, reverse osmosis, dialysis (biochemistry) utilising synthetic membranes, separates micrometre-sized or smaller species
- Fractional distillation
- Fractional freezing
- Oil-water separation, gravimetrically separates suspended oil droplets from waste water in oil refineries, petrochemical and chemical plants, natural gas processing plants and similar industries
- Magnetic separation
- Precipitation
- Recrystallization
- Scrubbing, separation of particulates (solids) or gases from a gas stream using liquid.

- Sedimentation, separates using vocal density pressure differences
    - Gravity separation
- Sieving
- Stripping
- Sublimation
- Vapor-liquid separation, separates by gravity, based on the Souders-Brown equation
- Winnowing
- Zone refining

REFERENCES

1. Wilson, Ian D.; Adlard, Edward R.; Cooke, Michael; et al., eds. (2000). Encyclopedia of separation science. San Diego: Academic Press. ISBN 978-0-12-226770-3.
2. *«Functional and constructional characteristics of separators».*

CHAPTER 2

Adsorption And Capillary Electrophoresis

Chapter Contents

- **Capillary Electrophoresis**
- **Applications**

INTRODUCTION

Adsorption is the adhesion of atoms, ions, or molecules from a gas, liquid, or dissolved solid to a surface. This process creates a film of the *adsorbate* on the surface of the *adsorbent.* This process differs from absorption, in which a fluid (the *absorbate*) is dissolved by or permeates a liquid or solid (the *absorbent*), respectively. Adsorption is a surface-based process while absorption involves the whole volume of the material. The term *sorption* encompasses both processes, while *desorption* is the reverse of it. Adsorption is a *surface phenomenon*.

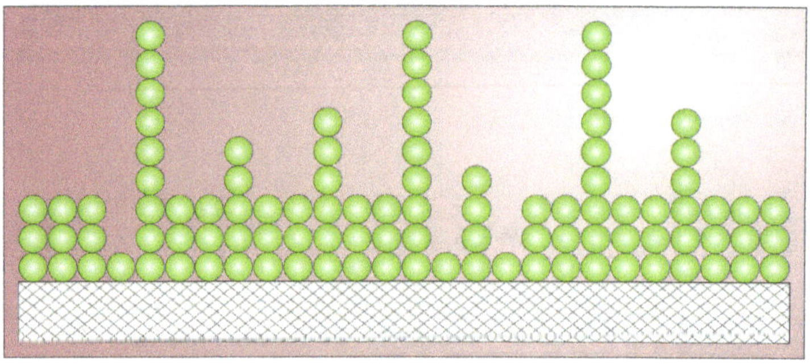

Brunauer, Emmett and Teller's model of multilayer adsorption is a random distribution of molecules on the material surface.

Similar to surface tension, adsorption is a consequence of surface energy. In a bulk material, all the bonding requirements (be they ionic, covalent, or metallic) of the constituent atoms of the material are filled by other atoms in the material. However, atoms on the surface of the adsorbent are not wholly surrounded by other adsorbent atoms and therefore can attract adsorbates. The exact nature of the bonding depends on the details of the species involved, but the adsorption process is generally classified as physisorption (characteristic of weak van der Waals forces) or chemisorption (characteristic of covalent bonding). It may also occur due to electrostatic attraction.

Adsorption is present in many natural, physical, biological, and chemical systems, and is widely used in industrial applications such as heterogeneous catalysts activated charcoal, capturing and using waste heat to provide cold water for air conditioning and other process requirements (adsorption chillers), synthetic resins, increase storage capacity of carbide-derived carbons, and water purification. Adsorption, ion exchange, and chromatography are sorption processes in which certain adsorbates are selectively transferred from the fluid phase to the surface of insoluble, rigid particles suspended in a vessel or packed in a column. Pharmaceutical industry applications, which use adsorption as a means to prolong neurological exposure to specific drugs or parts thereof, are lesser known.

The word «adsorption» was coined in 1881 by German physicist Heinrich Kayser (1853-1940).

Isotherms

Adsorption is usually described through isotherms, that is, the amount of adsorbate on the adsorbent as a function of its pressure (if gas) or concentration (if liquid) at constant temperature. The quantity adsorbed is nearly always normalized by the mass of the adsorbent to allow comparison of different materials. To date, 15 different isothem models were developed.

Freundlich

The first mathematical fit to an isotherm was published by Freundlich and Küster (1906) and is a purely empirical formula for gaseous adsorbates,

$$\frac{x}{m} = kP^{\frac{1}{n}}$$

where x is the quantity of adsorbate adsorbed in moles, m is the mass of the adsorbent, P is the pressure of adsorbate (this can be changed to concentration if investigating solution rather than gas) and k and n are empirical constants for each adsorbent-adsorbate pair at a given temperature. The function is not adequate at very high pressure because in reality x/m has an asymptotic maximum as pressure increases without bound. As the temperature increases, the constants k and n change to reflect the empirical observation that the quantity adsorbed rises more slowly and higher pressures are required to saturate the surface.

Langmuir

Irving Langmuir was the first to derive a scientifically based adsorption isotherm in 1918. The model applies to gases adsorbed on solid surfaces. It is a semi-empirical isotherm with a kinetic basis and was derived based on statistical thermodynamics. It is the most common isotherm equation to use due to its simplicity and its ability to fit a variety of adsorption data. It is based on four assumptions:

- All of the adsorption sites are equivalent and each site can only accommodate one molecule.
- The surface is energetically homogeneous and adsorbed molecules do not interact.
- There are no phase transitions.
- At the maximum adsorption, only a monolayer is formed. Adsorption only occurs on localized sites on the surface, not with other adsorbates.

These four assumptions are seldom all true: there are always imperfections on the surface, adsorbed molecules are not necessarily inert, and the mechanism is clearly not the same for the very first molecules to adsorb to a surface as for the last. The fourth condition is the most troublesome, as frequently more molecules will adsorb to the monolayer; this problem is addressed by the BET isotherm for relatively flat (non-microporous) surfaces. The Langmuir isotherm is nonetheless the first choice for most models of adsorption, and has many applications in surface kinetics (usually called Langmuir–Hinshelwood kinetics) and thermodynamics.

Langmuir suggested that adsorption takes place through this mechanism: $A_g + S \rightleftharpoons AS$, where A is a gas molecule and S is an adsorption site. The direct and inverse rate constants are k and k_{-1}. If we define surface coverage, θ, as the fraction of the adsorption sites occupied, in the equilibrium we have:

$$K = \frac{k}{k_{-1}} = \frac{\theta}{(1-\theta)P}$$

or

$$\theta = \frac{KP}{1+KP}$$

where P is the partial pressure of the gas or the molar concentration of the solution. For very low pressures $\theta \approx KP$ and for high pressures $\theta \approx 1$

θ is difficult to measure experimentally; usually, the adsorbate is a gas and the quantity adsorbed is given in moles, grams, or gas volumes at standard temperature and pressure (STP) per gram of adsorbent. If we call v_{mon} the STP volume of adsorbate required to form a monolayer on the adsorbent (per gram of adsorbent), $\theta = \frac{v}{v_{mon}}$ and we obtain an expression for a straight line:

$$\frac{1}{v} = \frac{1}{Kv_{mon}}\frac{1}{P} + \frac{1}{v_{mon}}$$

Through its slope and *y*-intercept we can obtain v_{mon} and *K*, which are constants for each adsorbent/adsorbate pair at a given temperature. v_{mon} is related to the number of adsorption sites through the ideal gas law. If we assume that the number of sites is just the whole area of the solid divided into the cross section of the adsorbate molecules, we can easily calculate the surface area of the adsorbent. The surface area of an adsorbent depends on its structure; the more pores it has, the greater the area, which has a big influence on reactions on surfaces.

If more than one gas adsorbs on the surface, we define θ_E as the fraction of empty sites and we have:

$$\theta_E = \frac{1}{1 + \sum_{i=1}^{n} K_i P_i}$$

Also, we can define θ_j as the fraction of the sites occupied by the *j*-th gas:

$$\theta_j = \frac{K_j P_j}{1 + \sum_{i=1}^{n} K_i P_i}$$

where *i* is each one of the gases that adsorb.

BET

Often molecules do form multilayers, that is, some are adsorbed on already adsorbed molecules and the Langmuir isotherm is not valid. In 1938 Stephen Brunauer, Paul Emmett, and Edward Teller developed a model isotherm that takes that possibility into account. Their theory is called BET theory, after the initials in their last names. They modified Langmuir›s mechanism as follows:

$A_{(g)} + S \rightleftharpoons AS$

$A_{(g)} + AS \rightleftharpoons A_2S$

$A_{(g)} + A_2S \rightleftharpoons A_3S$ and so on

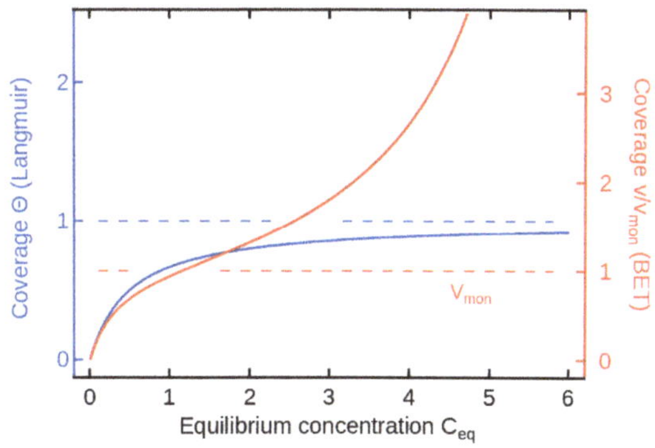

Langmuir (blue) and BET (red) isotherms

The derivation of the formula is more complicated than Langmuir's (see links for complete derivation). We obtain:

$$\frac{x}{v(1-x)} = \frac{1}{v_{\text{mon}}c} + \frac{x(c-1)}{v_{\text{mon}}c}.$$

x is the pressure divided by the vapor pressure for the adsorbate at that temperature (usually denoted P/P_0), v is the STP volume of adsorbed adsorbate, v_{mon} is the STP volume of the amount of adsorbate required to form a monolayer and c is the equilibrium constant K we used in Langmuir isotherm multiplied by the vapor pressure of the adsorbate. The key assumption used in deriving the BET equation that the successive heats of adsorption for all layers except the first are equal to the heat of condensation of the adsorbate.

The Langmuir isotherm is usually better for chemisorption and the BET isotherm works better for physisorption for non-microporous surfaces.

Kisliuk

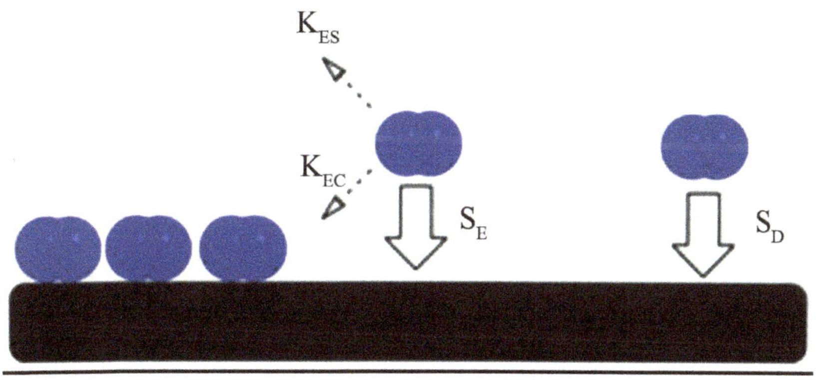

Two adsorbate nitrogen molecules adsorbing onto a tungsten adsorbent from the precursor state around an island of previously adsorbed adsorbate (left) and via random adsorption (right)

In other instances, molecular interactions between gas molecules previously adsorbed on a solid surface form significant interactions with gas molecules in the gaseous phases. Hence, adsorption of gas molecules to the surface is more likely to occur around gas molecules that are already present on the solid surface, rendering the Langmuir adsorption isotherm ineffective for the purposes of modelling. This effect was studied in a system where nitrogen was the adsorbate and tungsten was the adsorbent by Paul Kisliuk (1922–2008) in 1957. To compensate for the increased probability of adsorption occurring around molecules present on the substrate surface, Kisliuk developed the precursor state theory, whereby molecules would enter a precursor state at the interface between the solid adsorbent and adsorbate in the gaseous phase. From here, adsorbate molecules would either adsorb to the adsorbent or desorb into the gaseous phase. The probability of adsorption occurring from the precursor state is dependent on the adsorbate's proximity to other adsorbate molecules that have already been adsorbed. If the adsorbate molecule in the precursor state is in close proximity to an adsorbate molecule that has already formed on the surface, it has a sticking probability reflected by the size of the S_E constant and will either be adsorbed from the precursor state at a rate of k_{EC} or will desorb into the gaseous phase at a rate of k_{ES}. If an adsorbate molecule enters the precursor state at a location that is remote from any other previously adsorbed adsorbate molecules, the sticking probability is reflected by the size of the S_D constant.

These factors were included as part of a single constant termed a «sticking coefficient,» k_E, described below:

$$k_{\mathrm{E}} = \frac{S_{\mathrm{E}}}{k_{\mathrm{ES}} \, . \, S_{\mathrm{D}}}.$$

As S_D is dictated by factors that are taken into account by the Langmuir model, S_D can be assumed to be the adsorption rate constant. However, the rate constant for the Kisliuk model (R') is different from that of the Langmuir model, as R' is used to represent the impact of diffusion on monolayer formation and is proportional to the square root of the system's diffusion coefficient. The Kisliuk adsorption isotherm is written as follows, where $\Theta_{(t)}$ is fractional coverage of the adsorbent with adsorbate, and t is immersion time:

$$\frac{d\theta_{(\mathrm{t})}}{dt} = \mathrm{R}'(1-\theta)(1+k_{\mathrm{E}}\theta).$$

Solving for $\Theta_{(t)}$ yields:

$$\theta_{(\mathrm{t})} = \frac{1-e^{-R'(1+k_{\mathrm{E}})t}}{1+k_{\mathrm{E}}e^{-R'(1+k_{\mathrm{E}})t}}.$$

Adsorption Enthalpy

Adsorption constants are equilibrium constants, therefore they obey the van 't Hoff equation:

$$\left(\frac{\partial \ln K}{\partial \frac{1}{T}}\right)_{\theta} = -\frac{\Delta H}{R}.$$

As can be seen in the formula, the variation of K must be isosteric, that is, at constant coverage. If we start from the BET isotherm and assume that the entropy change is the same for liquefaction and adsorption we obtain

$$\Delta H_{\mathrm{ads}} = \Delta H_{\mathrm{liq}} - RT \ln c,$$

that is to say, adsorption is more exothermic than liquefaction.

Adsorbents

Characteristics and General Requirements

Activated carbon is used as an adsorbent

Adsorbents are used usually in the form of spherical pellets, rods, moldings, or monoliths with a hydrodynamic radius between 0.25 and 5 mm. They must have high abrasion resistance, high thermal stability and small pore diameters, which results in higher exposed surface area and hence high capacity for adsorption. The adsorbents must also have a distinct pore structure that enables fast transport of the gaseous vapors.

Most industrial adsorbents fall into one of three classes:

- Oxygen-containing compounds – Are typically hydrophilic and polar, including materials such as silica gel and zeolites.
- Carbon-based compounds – Are typically hydrophobic and non-polar, including materials such as activated carbon and graphite.
- Polymer-based compounds – Are polar or non-polar functional groups in a porous polymer matrix.

Silica Gel

Silica gel adsorber for NO_2, Fixed Nitrogen Research Laboratory, ca.1930s

Silica gel is a chemically inert, nontoxic, polar and dimensionally stable (< 400 °C or 750 °F) amorphous form of SiO_2. It is prepared by the reaction between sodium silicate and acetic acid, which is followed by a series of after-treatment processes such as aging, pickling, etc. These after-treatment methods results in various pore size distributions.

Silica is used for drying of process air (e.g. oxygen, natural gas) and adsorption of heavy (polar) hydrocarbons from natural gas.

Zeolites

Zeolites are natural or synthetic crystalline aluminosilicates, which have a repeating pore network and release water at high temperature. Zeolites are polar in nature.

They are manufactured by hydrothermal synthesis of sodium aluminosilicate or another silica source in an autoclave followed by ion exchange with certain cations (Na^+, Li^+, Ca^{2+}, K^+, NH_4^+). The channel diameter of zeolite cages usually ranges from 2 to 9 Å. The ion exchange process is followed by drying of the crystals, which can be pelletized with a binder to form macroporous pellets.

Zeolites are applied in drying of process air, CO_2 removal from natural gas, CO removal from reforming gas, air separation, catalytic cracking, and catalytic synthesis and reforming.

Non-polar (siliceous) zeolites are synthesized from aluminum-free silica sources or by dealumination of aluminum-containing zeolites. The dealumination process is done by treating the zeolite with steam at elevated temperatures, typically greater than 500 °C

(930 °F). This high temperature heat treatment breaks the aluminum-oxygen bonds and the aluminum atom is expelled from the zeolite framework.

Activated Carbon

Activated carbon is a highly porous, amorphous solid consisting of microcrystallites with a graphite lattice, usually prepared in small pellets or a powder. It is non-polar and cheap. One of its main drawbacks is that it reacts with oxygen at moderate temperatures (over 300 °C).

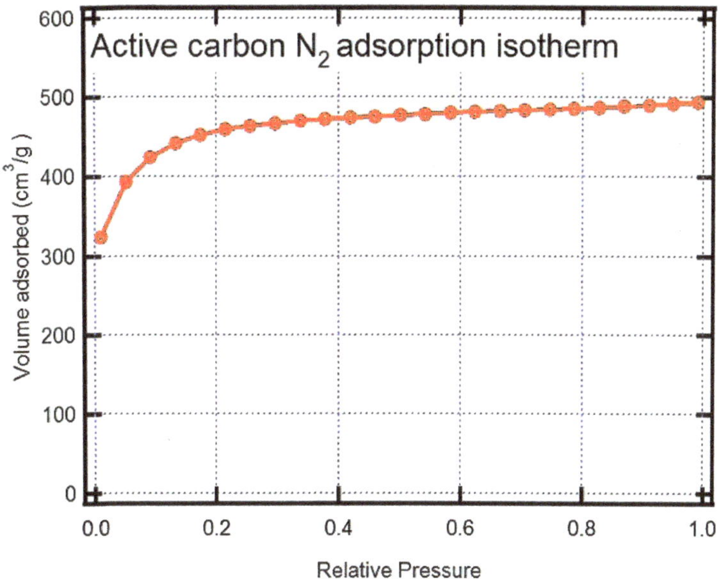

Activated carbon nitrogen isotherm showing a marked microporous type I behavior

Activated carbon can be manufactured from carbonaceous material, including coal (bituminous, subbituminous, and lignite), peat, wood, or nutshells (e.g., coconut). The manufacturing process consists of two phases, carbonization and activation. The carbonization process includes drying and then heating to separate by-products, including tars and other hydrocarbons from the raw material, as well as to drive off any gases generated. The process is completed by heating the material over 400 °C (750 °F) in an oxygen-free atmosphere that cannot support combustion. The carbonized particles are then "activated" by exposing them to an oxidizing agent, usually steam or carbon dioxide at high temperature. This agent burns off the pore blocking structures created during the carbonization phase and so, they develop a porous, three-dimensional graphite lattice structure. The size of the pores developed during activation is a function of the time that they spend in this stage. Longer exposure times result in larger pore sizes. The most popular aqueous phase carbons are bituminous based because of their hardness, abrasion resistance, pore size distribution, and low cost, but their effectiveness needs to be tested in each application to determine the optimal product.

Activated carbon is used for adsorption of organic substances and non-polar adsorbates and it is also usually used for waste gas (and waste water) treatment. It is the most widely used adsorbent since most of its chemical (e.g. surface groups) and physical properties (e.g. pore size distribution and surface area) can be tuned according to what is needed. Its usefulness also derives from its large micropore (and sometimes mesopore) volume and the resulting high surface area.

Protein Adsorption of Biomaterials

Protein adsorption is a process that has a fundamental role in the field of biomaterials. Indeed, biomaterial surfaces in contact with biological media, such as blood or serum, are immediately coated by proteins. Therefore, living cells do not interact directly with the biomaterial surface, but with the adsorbed proteins layer. This protein layer mediates the interaction between biomaterials and cells, translating biomaterial physical and chemical properties into a "biological language". In fact, cell membrane receptors bind to protein layer bioactive sites and these receptor-protein binding events are transduced, through the cell membrane, in a manner that stimulates specific intracellular processes that then determine cell adhesion, shape, growth and differentiation. Protein adsorption is influenced by many surface properties such as surface wettability, surface chemical composition and surface nanometre-scale morphology.

Adsorption Chillers

Combining an adsorbent with a refrigerant, adsorption chillers use heat to provide a cooling effect. This heat, in the form of hot water, may come from any number of industrial sources including waste heat from industrial processes, prime heat from solar thermal installations or from the exhaust or water jacket heat of a piston engine or turbine.

Although there are similarities between absorption and adsorption refrigeration, the latter is based on the interaction between gases and solids. The adsorption chamber of the chiller is filled with a solid material (for example zeolite, silica gel, alumina, active carbon and certain types of metal salts), which in its neutral state has adsorbed the refrigerant. When heated, the solid desorbs (releases) refrigerant vapour, which subsequently is cooled and liquefied. This liquid refrigerant then provides its cooling effect at the evaporator, by *ab*sorbing external heat and turning back into a vapour. In the final stage the refrigerant vapour is (re) adsorbed into the solid. As an adsorption chiller requires no moving parts, it is relatively quiet.

Portal Site Mediated Adsorption

Portal site mediated adsorption is a model for site-selective activated gas adsorption in metallic catalytic systems that contain a variety of different adsorption sites. In such systems, low-coordination "edge and corner" defect-like sites can exhibit significantly lower adsorption enthalpies than high-coordination (basal plane) sites. As a result, these sites can serve as "portals" for very rapid adsorption to the rest of the surface. The phenomenon relies on the common "spillover" effect (described below), where certain adsorbed species exhibit high mobility on some surfaces. The model explains seemingly inconsistent observations of gas adsorption thermodynamics and kinetics in catalytic systems where surfaces can exist in a range of coordination structures, and it has been successfully applied to bimetallic catalytic systems where synergistic activity is observed.

In contrast to pure spillover, portal site adsorption refers to surface diffusion to adjacent adsorption sites, not to non-adsorptive support surfaces.

The model appears to have been first proposed for carbon monoxide on silica-supported platinum by Brandt *et al.* (1993). A similar, but independent model was developed by King and co-workers to describe hydrogen adsorption on silica-supported alkali promoted ruthenium, silver-ruthenium and copper-ruthenium bimetallic catalysts. The same group applied the model to CO hydrogenation (Fischer–Tropsch synthesis). Zupanc *et al.* (2002) subsequently confirmed the same model for hydrogen adsorption on magnesia-supported caesium-ruthenium bimetallic catalysts. Trens *et al.* (2009) have similarly described CO surface diffusion on carbon-supported Pt particles of varying morphology.

Adsorption Spillover

In the case catalytic or adsorbent systems where a metal species is dispersed upon a support (or carrier) material (often quasi-inert oxides, such as alumina or silica), it is possible for an adsorptive species to indirectly adsorb to the support surface under conditions where such adsorption is thermodynamically unfavorable. The presence of the metal serves as a lower-energy pathway for gaseous species to first adsorb to the metal and then diffuse on the support surface. This is possible because the adsorbed species attains a lower energy state once it has adsorbed to the metal, thus lowering the activation barrier between the gas phase species and the support-adsorbed species.

Hydrogen spillover is the most common example of an adsorptive spillover. In the case of hydrogen, adsorption is most often accompanied with dissociation of molecular hydrogen (H_2) to atomic hydrogen (H), followed by spillover of the hydrogen atoms present.

The spillover effect has been used to explain many observations in heterogeneous catalysis and adsorption.

Polymer Adsorption

Adsorption of molecules onto polymer surfaces is central to a number of applications, including development of non-stick coatings and in various biomedical devices. Polymers may also be adsorbed to surfaces through polyelectrolyte adsorption.

Adsorption in Viruses

Adsorption is the first step in the viral life cycle. The next steps are penetration, uncoating, synthesis (transcription if needed, and translation), and release. The virus replication cycle, in this respect, is similar for all types of viruses. Factors such as transcription may or may not be needed if the virus is able to integrate its genomic information in the cell›s nucleus, or if the virus can replicate itself directly within the cell›s cytoplasm.

In Popular Culture

The game of Tetris is a puzzle game in which blocks of 4 are adsorbed onto a surface during game play. Scientists have used Tetris blocks "as a proxy for molecules with a complex shape" and their "adsorption on a flat surface" for studying the thermodynamics of nanoparticles.

CAPILLARY ELECTROPHORESIS

Capillary electrophoresis (CE) is a family of electrokinetic separation methods performed in submillimeter diameter capillaries and in micro- and nanofluidic channels. Very often, CE refers to capillary zone electrophoresis (CZE), but other electrophoretic techniques including capillary gel electrophoresis (CGE), capillary isoelectric focusing (CIEF), capillary isotachophoresis and micellar electrokinetic chromatography (MEKC) belong also to this class of methods. In CE methods, analytes migrate through electrolyte solutions under the influence of an electric field. Analytes can be separated according to ionic mobility and/or partitioning into an alternate phase via Non-covalent interactions. Additionally, analytes may be concentrated or «focused» by means of gradients in conductivity and pH.

Instrumentation

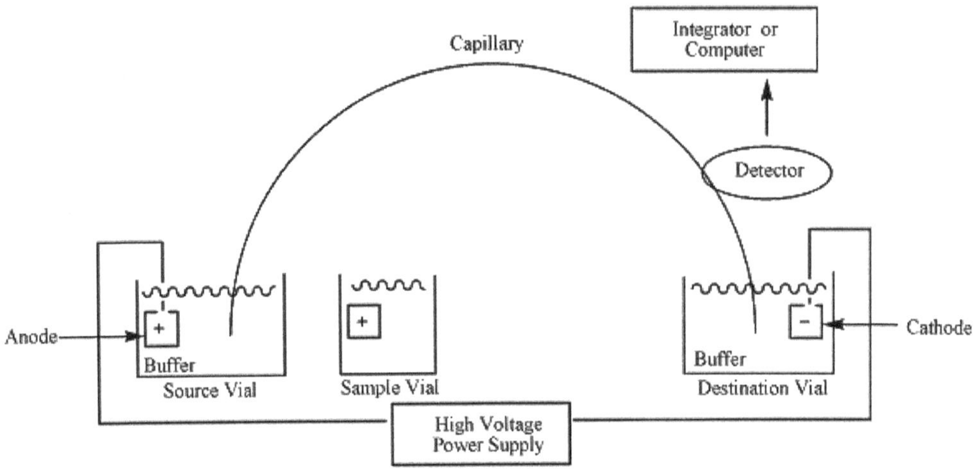

Figure 1: Diagram of capillary electrophoresis system.

The instrumentation needed to perform capillary electrophoresis is relatively simple. A basic schematic of a capillary electrophoresis system is shown in *figure 1*. The system's main components are a sample vial, source and destination vials, a capillary, electrodes, a high voltage power supply, a detector, and a data output and handling device. The source vial, destination vial and capillary are filled with an electrolyte such as an aqueous buffer solution. To introduce the sample, the capillary inlet is placed into a vial containing the sample. Sample is introduced into the capillary via capillary action, pressure, siphoning, or electrokinetically, and the capillary is then returned to the source vial. The migration of the analytes is initiated by an electric field that is applied between the source and destination vials and is supplied to the electrodes by the high-voltage power supply. In the most common mode of CE, all ions, positive or negative, are pulled through the capillary in the same direction by electroosmotic flow. The analytes separate as they migrate due to their electrophoretic mobility, and are detected near the outlet end of the capillary. The output of the detector is sent to a data output and handling device such as an integrator or computer. The data is then displayed as an electropherogram, which reports detector response as a function of time. Separated chemical compounds appear as peaks with different retention times in an electropherogram. Capillary electrophoresis was first combined with mass spectrometry by Richard D. Smith and coworkers, and provides extremely high sensitivity for the analysis of very small sample sizes. Despite the very small sample sizes (typically only a few nanoliters of liquid are introduced into the capillary), high sensitivity and sharp peaks are achieved in part due to injection strategies that result in concentration of analytes into a narrow zone near the inlet of the capillary. This is achieved in either pressure or electrokinetic injections simply by suspending the sample in a buffer of lower conductivity (*e.g.* lower salt concentration) than the running buffer. A process called field-amplified sample stacking (a form of isotachophoresis) results in concentration of analyte in a narrow zone at the boundary between the low-conductivity sample and the higher-conductivity running buffer.

To achieve greater sample throughput, instruments with arrays of capillaries are used to analyze many samples simultaneously. Such capillary array electrophoresis (CAE) instruments with 16 or 96 capillaries are used for medium- to high-throughput capillary DNA sequencing, and the inlet ends of the capillaries are arrayed spatially to accept samples directly from SBS-standard footprint 96-well plates. Certain aspects of the instrumentation (such as detection) are necessarily more complex than for a single-capillary system, but the fundamental principles of design and operation are similar to those shown in Figure 1.

Detection

Separation by capillary electrophoresis can be detected by several detection devices. The majority of commercial systems use UV or UV-Vis absorbance as their primary mode of detection. In these systems, a section of the capillary itself is used as the detection cell. The use of on-tube detection enables detection of separated analytes with no loss of resolution. In general, capillaries used in capillary electrophoresis are coated with a polymer (frequently polyimide or Teflon) for increased flexibility. The portion of the capillary used for UV detection, however, must be optically transparent. For polyimide-coated capillaries, a segment of the coating is typically burned or scraped off to provide a bare window several millimeters long. This bare section of capillary can break easily, and capillaries with transparent coatings are available to increase the stability of the cell window. The path length of the detection cell in capillary electrophoresis (~ 50 micrometers) is far less than that of a traditional UV cell (~ 1 cm). According to the Beer-Lambert law, the sensitivity of the detector is proportional to the path length of the cell. To improve the sensitivity, the path length can be increased, though this results in a loss of resolution. The capillary tube itself can be expanded at the detection point, creating a «bubble cell» with a longer path length or additional tubing can be added at the detection point as shown in *figure 2*. Both of these methods, however, will decrease the resolution of the separation. Post-column detection utilizing a sheath flow configuration has also been described.

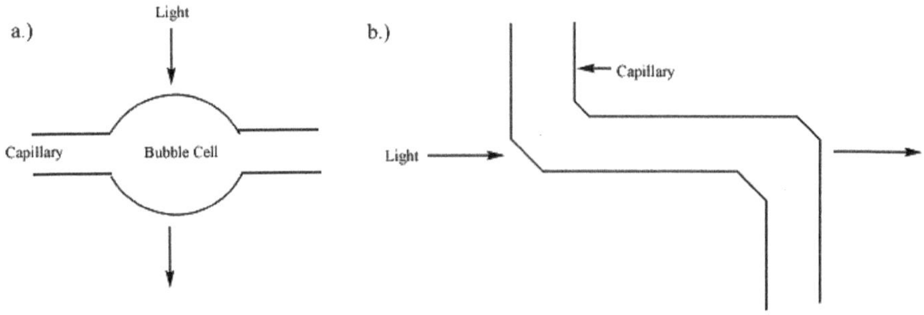

Figure 2: Techniques for increasing the pathlength of the capillary: a.) a bubble cell and b.) a z-cell (additional tubing).

Fluorescence detection can also be used in capillary electrophoresis for samples that naturally fluoresce or are chemically modified to contain fluorescent tags. This mode of detection offers high sensitivity and improved selectivity for these samples, but cannot be utilized for samples that do not fluoresce. Numerous labeling strategies are used to create fluorescent derivatives or conjugates of non-fluorescent molecules, including proteins and DNA. The set-up for fluorescence detection in a capillary electrophoresis system can be complicated. The method requires that the light beam be focused on the capillary, which can be difficult for many light sources. Laser-induced fluorescence has been used in CE systems with detection limits as low as 10^{-18} to 10^{-21} mol. The sensitivity of the technique is attributed to the high intensity of the incident light and the ability to accurately focus the light on the capillary. Multi-color fluorescence detection can be achieved by including multiple dichroic mirrors and bandpass filters to separate the fluorescence emission amongst multiple detectors (*e.g.*, photomultiplier tubes), or by using a prism or grating to project spectrally resolved fluorescence emission onto a position-sensitive detector such as a CCD array. CE systems with 4- and 5-color LIF detection systems are used routinely for capillary DNA sequencing and genotyping ("DNA fingerprinting») applications.

In order to obtain the identity of sample components, capillary electrophoresis can be directly coupled with mass spectrometers or Surface Enhanced Raman Spectroscopy (SERS). In most systems, the capillary outlet is introduced into an ion source that utilizes electrospray ionization (ESI). The resulting ions are then analyzed by the mass spectrometer. This set-up requires volatile buffer solutions, which will affect the range of separation modes that can be employed and the degree of resolution that can be achieved. The measurement and analysis are mostly done with a specialized gel analysis software.

For CE-SERS, capillary electrophoresis eluants can be deposited onto a SERS-active substrate. Analyte retention times can be translated into spatial distance by moving the SERS-active substrate at a constant rate during capillary electrophoresis. This allows the subsequent spectroscopic technique to be applied to specific eluants for identification with high sensitivity. SERS-active substrates can be chosen that do not interfere with the spectrum of the analytes.

Modes of Separation

The separation of compounds by capillary electrophoresis is dependent on the differential migration of analytes in an applied electric field. The electrophoretic migration velocity (u_p) of an analyte toward the electrode of opposite charge is:

$$u_p = \mu_p E$$

The electrophoretic mobility can be determined experimentally from the migration time and the field strength:

$$\mu_p = \left(\frac{L}{t_r}\right)\left(\frac{L_t}{V}\right)$$

where L is the distance from the inlet to the detection point, t_r is the time required for the analyte to reach the detection point (migration time), V is the applied voltage (field strength), and L_t is the total length of the capillary. Since only charged ions are affected by the electric field, neutral analytes are poorly separated by capillary electrophoresis.

The velocity of migration of an analyte in capillary electrophoresis will also depend upon the rate of electroosmotic flow (EOF) of the buffer solution. In a typical system, the electroosmotic flow is directed toward the negatively charged cathode so that the buffer flows through the capillary from the source vial to the destination vial. Separated by differing electrophoretic mobilities, analytes migrate toward the electrode of opposite charge. As a result, negatively charged analytes are attracted to the positively charged anode, counter to the EOF, while positively charged analytes are attracted to the cathode, in agreement with the EOF as depicted in *figure 3*.

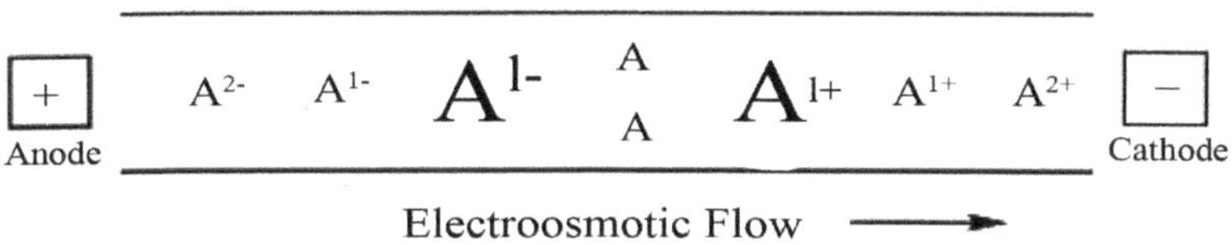

Figure 3: Diagram of the separation of charged and neutral analytes (A) according to their respective electrophoretic and electroosmotic flow mobilities.

The velocity of the electroosmotic flow, u_o can be written as:

$$u_o = \mu_o E$$

where μ_o is the electroosmotic mobility, which is defined as:

$$\mu_o = \frac{\epsilon\zeta}{\eta}$$

where ζ is the zeta potential of the capillary wall, and ϵ is the relative permittivity of the buffer solution. Experimentally, the electroosmotic mobility can be determined by measuring the retention time of a neutral analyte.The velocity u of an analyte in an electric field can then be defined as:

$$u_p + u_o = (\mu_p + \mu_o)E$$

Since the electroosmotic flow of the buffer solution is generally greater than that of the electrophoretic mobility of the analytes, all analytes are carried along with the buffer solution toward the cathode. Even small, triply charged anions can be redirected to the cathode by the relatively powerful EOF of the buffer solution. Negatively charged analytes are retained longer in the capilliary due to their conflicting electrophoretic mobilities. The order of migration seen by the detector is shown in *figure 3*: small multiply charged cations migrate quickly and small multiply charged anions are retained strongly.

Electroosmotic flow is observed when an electric field is applied to a solution in a capillary that has fixed charges on its interior wall. Charge is accumulated on the inner surface of a capillary when a buffer solution is placed inside the capillary. In a fused-silica capillary, silanol (Si-OH) groups attached to the interior wall of the capillary are ionized to negatively charged silanoate ($Si-O^-$) groups at pH values greater than three. The ionization of the capillary wall can be enhanced by first running a basic solution, such as NaOH or KOH through the capillary prior to introducing the buffer solution. Attracted to the negatively charged silanoate groups, the positively charged cations of the buffer solution will form two inner layers of cations (called the diffuse double layer or the electrical double layer) on the capillary wall as shown in *figure 4*. The first layer is referred to as the fixed layer because it is held tightly to the silanoate groups. The outer layer, called the mobile layer, is farther from the silanoate groups. The mobile cation layer is pulled in the direction of the negatively charged cathode when an electric field is applied. Since these cations are solvated, the bulk buffer solution migrates with the mobile layer, causing the electroosmotic flow of the buffer solution. Other capillaries including Teflon capillaries also exhibit electroosmotic flow. The EOF of these capillaries is probably the result of adsorption of the electrically charged ions of the buffer onto the capillary walls. The rate of EOF is dependent on the field strength and the charge density of the capillary wall. The wall's charge density is proportional to the pH of the buffer solution. The electroosmotic flow will increase with pH until all of the available silanols lining the wall of the capillary are fully ionized.

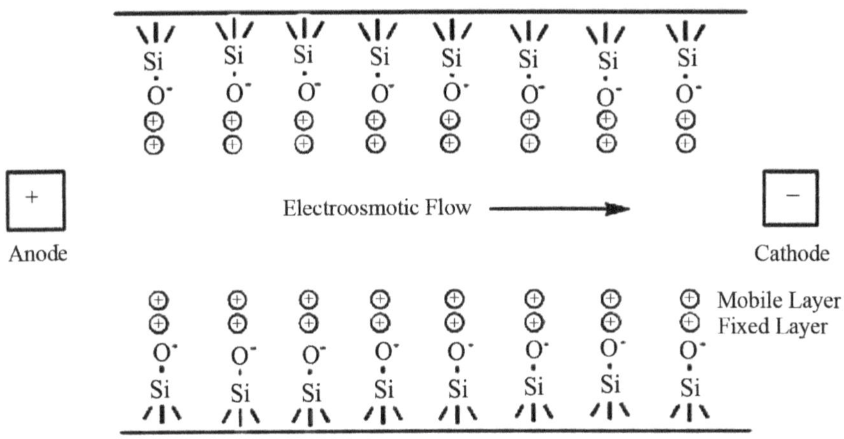

Figure 4: Depiction of the interior of a fused-silica gel capillary in the presence of a buffer solution.

In certain situations where strong electroosomotic flow toward the cathode is undesirable, the inner surface of the capillary can be coated with polymers, surfactants, or small molecules to reduce electroosmosis to very low levels, restoring the normal direction of migration (anions towar the anode, cations toward the cathode). CE instrumentation typically includes power supplies with reversible polarity, allowing the same instrument to be used in "normal" mode (with EOF and detection near the cathodic end of the capillary) and "reverse" mode (with EOF suppressed or reversed, and detection near the anodic end of the capillary). One of the most common approaches to suppressing EOF, reported by Stellan Hjertén in 1985, is to create a covalently attached layer of linear polyacrylamide. The silica surface of the capillary is first modified with a silane reagent bearing a polymerizable vinyl group (*e.g.* 3-methacryloxypropyltrimethoxys ilane), followed by introduction of acrylamide monomer and a free radical initiator. The acrylamide is polymerized *in situ*, forming long linear chains, some of which are covalently attached to the wall-bound silane reagent. Numerous other strategies for covalent modification of capillary surfaces exist. Dynamic or adsorbed coatings (which can include polymers or small molecules) are also common. For example, in capillary sequencing of DNA, the sieving polymer (typically polydimethylacrylamide) suppresses electroosmotic flow to very low levels. A variety of dynamic capillary coating agents are commercially available to modify, suppress, or reverse the direction of electroosmotic flow. Besides modulating electroosmotic flow, capillary wall coatings can also serve the purpose of reducing interactions between "sticky" analytes (such as proteins) and the capillary wall. Such wall-analyte interactions, if severe, manifest as reduced peak efficiency, asymmetric (tailing) peaks, or even complete loss of analyte to the capillary wall.

Efficiency and Resolution

The number of theoretical plates, or separation efficiency, in capillary electrophoresis is given by:

$$N = \frac{\mu V}{2D_m}$$

where N is the number of theoretical plates, μ is the apparent mobility in the separation medium and D_m is the diffusion coefficient of the analyte. According to this equation, the efficiency of separation is only limited by diffusion and is proportional to the strength of the electric field, although practical considerations limit the strength of the electric field to several hundred volts per centimeter. Application of very high potentials (>20-30 kV) may lead to arcing or breakdown of the capillary. Further, application of strong electric fields leads to resistive heating (Joule heating) of the buffer in the capillary. At sufficiently high field strengths, this heating is strong enough that radial temperature gradients can develop within the capillary. Since electrophoretic mobility of ions is generally temperature-dependent (due to both temperature-dependent ionization and solvent viscosity effects), a non-uniform temperature profile results in variation of electrophoretic mobility across the capillary, and a loss of resolution. The onset of significant Joule heating can be determined by constructing an «Ohm›s Law plot», wherein the current through the capillary is measured as a function of applied potential. At low fields, the current is proportional to the applied potential (Ohm›s Law), whereas at higher fields the current deviates from the straight line as heating results in decreased resistance of the buffer. The best resolution is typically obtained at the maximum field strength for which Joule heating is insignificant (*i.e.* near the boundary between the linear and nonlinear regimes of the Ohm's Law plot). Generally capillaries of smaller inner diameter support use of higher field strengths, due to improved heat dissipation and smaller thermal gradients relative to larger capillaries, but with the drawbacks of lower sensitivity in absorbance detection due to shorter path length, and greater difficulty in introducing buffer and sample into the capillary (small capillaries require greater pressure and/or longer times to force fluids through the capillary).

The efficiency of capillary electrophoresis separations is typically much higher than the efficiency of other separation techniques like HPLC. Unlike HPLC, in capillary electrophoresis there is no mass transfer between phases. In addition, the flow profile in EOF-driven systems is flat, rather than the rounded laminar flow profile characteristic of the pressure-driven flow in chromatography columns as shown in *figure 5*. As a result, EOF does not significantly contribute to band broadening as in pressure-driven chromatography. Capillary electrophoresis separations can have several hundred thousand theoretical plates.

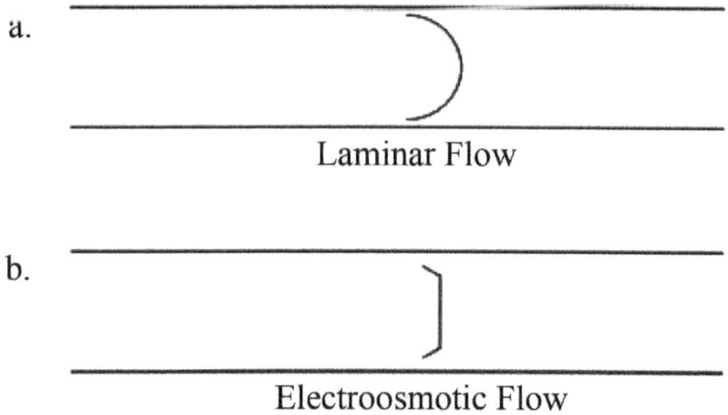

Figure 5: Flow profiles of laminar and electroosmotic flow.

The resolution (R_s) of capillary electrophoresis separations can be written as:

$$R_s = \frac{1}{4}\left(\frac{\Delta\mu_p\sqrt{N}}{\mu_p + \mu_o}\right)$$

According to this equation, maximum resolution is reached when the electrophoretic and electroosmotic mobilities are similar in magnitude and opposite in sign. In addition, it can be seen that high resolution requires lower velocity and, correspondingly, increased analysis time.

Besides diffusion and Joule heating (discussed above), factors that may decrease the resolution in capillary electrophoresis from the theoretical limits in the above equation include, but are not limited to, the finite widths of the injection plug and detection window; interactions between the analyte and the capillary wall; instrumental non-idealities such as a slight difference in height of the fluid reservoirs leading to siphoning; irregularities in the electric field due to, *e.g.*, imperfectly cut capillary ends; depletion of buffering capacity in the reservoirs; and electrodispersion (when an analyte has higher conductivity than the background electrolyte). Identifying and minimizing the numerous sources of band broadening is key to successful method development in capillary electrophoresis, with the objective of approaching as close as possible to the ideal of diffusion-limited resolution.

APPLICATIONS

Capillary electrophoresis may be used for the simultaneous determination of the ions NH_4^+, Na^+, K^+, Mg^{2+} and Ca^{2+} in saliva.

REFERENCES

1. "Glossary". The Brownfields and Land Revitalization Technology Support Center. Retrieved 2009-12-21.
2. "absorption (chemistry)". Memidex (WordNet) Dictionary/Thesaurus. Retrieved 2010-11-02.
3. "Glossary of atmospheric chemistry terms (Recommendations 1990)". Pure and Applied Chemistry. 62: 2167. 1990. doi:10.1351/goldbook.A00155. ISBN 0-9678550-9-8.
4. Ferrari, L.; Kaufmann, J.; Winnefeld, F.; Plank, J. (2010). "Interaction of cement model systems with superplasticizers investigated by atomic force microscopy, zeta potential, and adsorption measurements". J Colloid Interface Sci. 347 (1): 15–24. doi:10.1016/j.jcis.2010.03.005. PMID 20356605.
5. 2.Czelej, K., Cwieka, K., Kurzydlowski, K.J., CO2 stability on the Ni low-index surfaces: Van der Waals corrected DFT analysis, Catalysis Communications 2016, 80 p. 33-38.
6. 1.Czelej K., Cwieka K., Colmenares J.C., Kurzydlowski K.J. Insight on the Interaction of Methanol-Selective Oxidation Intermediates with Au- or/and Pd-Containing Monometallic and Bimetallic Core@Shell Catalysts, Langmuir 2016 32 (30), p. 7493-7502
7. Kayser, Heinrich (1881). "Über die Verdichtung von Gasen an Oberflächen in ihrer Abhängigkeit von Druck und Temperatur". Annalen der Physik und Chemie. 248 (4): 526–537. Bibcode:1881AnP...248..526K. doi:10.1002/andp.18812480404.. In this study of the adsorption of gases by charcoal, the first use of the word "adsorption" appears on page 527: "Schon Saussure kannte die beiden für die Grösse der Adsorption massgebenden Factoren, den Druck und die Temperatur, da er Erniedrigung des Druckes oder Erhöhung der Temperatur zur Befreiung der porösen Körper von Gasen benutzte." (Saussaure already knew the two factors that determine the quantity of adsorption – [namely,] the pressure and temperature – since he used the lowering of the pressure or the raising of the temperature to free the porous substances of gases.)
8. Foo, K.Y.; Hameed, B.H. (2010). "Insights into the modeling of adsorption isotherm systems". Chemical Engineering Journal. 156 (1): 2–10. doi:10.1016/j.cej.2009.09.013. ISSN 1385-8947.
9. Czepirski, L.; Balys, M. R.; Komorowska-Czepirska, E. (2000). "Some generalization of Langmuir adsorption isotherm". Internet Journal of Chemistry. 3 (14). ISSN 1099-8292.
10. Kisliuk, P. (1957). "The sticking probabilities of gases chemisorbed on the surfaces of solids". Journal of Physics and Chemistry of Solids. 3 (1–2): 95–101. Bibcode:1957JPCS....3...95K. doi:10.1016/0022-3697(57)90054-9.
11. Wilson, CJ; Clegg, RE; Leavesley, DI; Pearcy, MJ (2005). "Mediation of Biomaterial-Cell Interactions by Adsorbed Proteins: A Review". Tissue engineering. 11 (1): 1–18. doi:10.1089/ten.2005.11.1. PMID 15738657.

12. Sivaraman B.; Fears K.P.; Latour R.A. (2009). "Investigation of the effects of surface chemistry and solution concentration on the conformation of adsorbed proteins using an improved circular dichroism method". Langmuir. 25 (5): 3050–6. doi:10.1021/la8036814. PMC 2891683 . PMID 19437712.

13. Scopelliti, Pasquale Emanuele; Borgonovo, Antonio; Indrieri, Marco; Giorgetti, Luca; Bongiorno, Gero; Carbone, Roberta; Podestà, Alessandro; Milani, Paolo (2010). Zhang, Shuguang, ed. "The effect of surface nanometre-scale morphology on protein adsorption". PLoS ONE. 5 (7): e11862. Bibcode:2010PLoSO...511862S. doi:10.1371/journal.pone.0011862.

14. Pilatowsky, I.; Romero, R.J.; Isaza, C.A.; Gamboa, S.A.; Sebastian, P.J.; Rivera, W. (2011). "Chapter 5: Sorption Refrigeration Systems". Cogeneration Fuel Cell-Sorption Air Conditioning Systems. Green Energy and Technology. Springer. pp. 99,100. ISBN 978-1-84996-027-4. Retrieved 10 May 2011.

15. Brandt, R. K.; Hughes, M. R.; Bourget, L. P.; Truszkowska, K.; Greenler, R. G. (1993). "The interpretation of CO adsorbed on Pt/SiO2of two different particle-size distributions". Surface Science. 286 (1–2): 15–25. Bibcode:1993SurSc.286...15B. doi:10.1016/0039-6028(93)90552-U.

16. Uner, D. O.; Savargoankar, N.; Pruski, M.; King, T. S. (1997). "The effects of alkali promoters on the dynamics of hydrogen chemisorption and syngas reaction kinetics on Ru/SiO2 catalysts". Studies in Surface Science and Catalysis. Studies in Surface Science and Catalysis. 109: 315–324. doi:10.1016/S0167-2991(97)80418-1. ISBN 9780444826091.

17. Narayan, R. L.; King, T. S. (1998). "Hydrogen adsorption states on silica-supported Ru-Ag and Ru-Cu bimetallic catalysts investigated via microcalorimetry". Thermochimica Acta. 312 (1–2): 105–114. doi:10.1016/S0040-6031(97)00444-9.

18. VanderWiel, D. P.; Pruski, M.; King, T. S. (1999). "A Kinetic Study of the Adsorption and Reaction of Hydrogen on Silica-Supported Ruthenium and Silver-Ruthenium Bimetallic Catalysts during the Hydrogenation of Carbon Monoxide". Journal of Catalysis. 188 (1): 186–202. doi:10.1006/jcat.1999.2646.

19. Uner, D. O. (1998). "A sensible mechanism of alkali promotion in Fischer Tropsch synthesis:Adsorbate mobilities". Industrial and Engineering Chemistry Research. 37 (6): 2239–2245. doi:10.1021/ie970696d.

20. Zupanc, C.; Hornung, A.; Hinrichsen, O.; Muhler, M. (2002). "The Interaction of Hydrogen with Ru/MgO Catalysts". Journal of Catalysis. 209 (2): 501–514. doi:10.1006/jcat.2002.3647.

21. Trens, P.; Durand, R.; Coq, B.; Coutanceau, C.; Rousseau, S.; Lamy, C. (2009). "Poisoning of Pt/C catalysts by CO and its consequences over the kinetics of hydrogen chemisorption". Applied Catalysis B: Environmental. 92 (3–4): 280–284. doi:10.1016/j.apcatb.2009.08.004.

22. Rozanov, V. V.; Krylov, O. V. (1997). "Hydrogen spillover in heterogeneous catalysis". Russian Chemical Reviews. 66 (2): 107–119. Bibcode:1997RuCRv..66..107R. doi:10.1070/RC1997v066n02ABEH000308.

23. The Thermodynamics of Tetiris, Ars Technica, 2009.

24. Barnes, Brian C.; Siderius, Daniel W.; Gelb, Lev D. (2009). "Structure, Thermodynamics, and Solubility in Tetromino Fluids". Langmuir. 25 (12): 6702–16. doi:10.1021/la900196b. PMID 19397254.
25. Graham Kemp, Capillary electrophoresis: a versatile family of analytical techniques Archived April 27, 2006, at the Wayback Machine. Biotechnology and Applied Biochemistry (1998) 27, (9–17)
26. Skoog, D.A.; Holler, F.J.; Crouch, S.R "Principles of Instrumental Analysis" 6th ed. Thomson Brooks/Cole Publishing: Belmont, CA 2007.
27. Skoog, D.A.; Holler, F.J.; Crouch, S.R "Principles of Instrumental Analysis" 6th ed. Chapter 30 Thomson Brooks/Cole Publishing: Belmont, CA 2007.
28. Dovichi, Norman (2000). "How capillary electrophoresis sequenced the human genome" (PDF). Angewandte Chemie International Edition. 39: 4463–4468. doi:10.1002/1521-3773(20001215)39:24<4463::aid-anie4463>3.0.co;2-8. Retrieved 2014-04-09.
29. Butler, John (2004). "Forensic DNA typing by capillary electrophoresis using the ABI Prism 310 and 3100 genetic analyzers for STR analysis" (PDF). Electrophoresis. 25: 1397–1412. doi:10.1002/elps.200305822. PMID 15188225. Retrieved 2014-04-09.
30. Lin H.; Natan, M.; Keating, C. Anal. Chem. 2000, 72, 5348-5355.
31. Hjertén, Stellan (1985). "High-performance electrophoresis: elimination of electroosmosis and solute adsorption". Journal of Chromatography (347): 191–198.
32. Doherty EA, Meagher RJ, Albarghouthi MN, Barron AE (2003). "Microchannel wall coatings for protein separations by capillary and chip electrophoresis". Electrophoresis. 24: 34–54. doi:10.1002/elps.200390029. PMID 12652571.
33. Mandabhushi, Ramakrishna (1998). "Separation of 4-color DNA sequencing extension products in noncovalently coated capillaries using low viscosity polymer solutions". Electrophoresis. 19: 224–230. doi:10.1002/elps.1150190215. PMID 9548284.
34. Target Discovery. "UltraTrol™ Dynamic Pre-Coatings". Retrieved 9 April 2014.
35. Skoog, D.A.; Holler, F.J.; Nieman, T.A. "Principles of Instrumental Analysis, 5th ed." Saunders college Publishing: Philadelphia, 1998.
36. Lauer and Rozing. "High Performance Capillary Electrophoresis: A primer" (PDF). Agilent Technologies. Archived from the original (PDF) on April 13, 2014. Retrieved 2014-04-09.
37. Hauser, Peter C. (2016). "Chapter 2. Determination of Alkali Ions in Biological and Environmental Samples". In Astrid, Sigel; Helmut, Sigel; Roland K.O., Sigel. The Alkali Metal Ions: Their Role in Life. Metal Ions in Life Sciences. 16. Springer. pp. 11–25. doi:10.1007/978-3-319-21756-7_2.

CHAPTER 3

Centrifugation and Cyclonic Separation

Chapter Contents

- **Introduction**
- **Cyclonic Separation**
- **Chromatography**
- **History**
- **Chromatography Terms**
- **Displacement Chromatography**
- **Techniques By Separation Mechanism**

Centrifugation is the process by which a centrifuge is used to separate components of a complex mixture. By spinning laboratory samples at very high speeds, the components of a given mixture are subjected to centrifugal force, which causes more dense particles to migrate away from the axis of rotation and lighter ones to move toward it. These particles can sediment at the bottom of the tube into what's known as a pellet, and this isolated specimen, or the remaining solution, the supernatant, can be further processed or analyzed. This video is meant to introduce a student to some of the basic principles of centrifugation, as well as the instrument's basic operation. For example the speed of centrifugation in rotations per minute, or RPM, is contrasted with relative centrifugal force, or RCF, as a measurement of the magnitude of centrifugation, which is independent of rotor size. In addition to concepts and basic use, safety precautions relating to centrifugation are discussed, as well as the types of centrifuges and centrifuge rotors that exist.

INTRODUCTION

Centrifugation is a process which involves the application of the centrifugal force for the sedimentation of heterogeneous mixtures with a centrifuge, and is used in industrial and laboratory settings. This process is used to separate two miscible substances, but also to analyze the hydrodynamic properties of macromolecules. More-dense components of the mixture migrate away from the axis of the centrifuge, while less-dense components of the mixture migrate towards the axis. Chemists and biologists may increase the effective gravitational force on a test tube so as to more rapidly and completely cause the precipitate (pellet) to gather on the bottom of the tube. The remaining solution (supernatant) may be discarded with a pipette.

There is a correlation between the size and density of a particle and the rate that the particle separates from a heterogeneous mixture, when the only force applied is that of gravity. The larger the size and the larger the density of the particles, the faster they separate from the mixture. By applying a larger effective gravitational force to the mixture, like a centrifuge does, the separation of the particles is accelerated. This is ideal in industrial and lab settings because particles that would naturally separate over a long period of time can be separated in much less time.

The rate of centrifugation is specified by the angular velocity usually expressed as revolutions per minute (RPM), or acceleration expressed as *g*. The conversion factor between RPM and *g* depends on the radius of the centrifuge rotor. The particles› settling velocity in centrifugation is a function of their size and shape, centrifugal acceleration, the volume fraction of solids present, the density difference between the particle and the liquid, and the viscosity. The most common application is the separation of solid from highly concentrated suspensions, which is used in the treatment of sewage sludges for dewatering where less consistent sediment is produced.

In the chemical and food industries, special centrifuges can process a continuous stream of particle-laden liquid.

Centrifugation is the most common method used for uranium enrichment, relying on the slight mass difference between atoms of U238 and U235 in uranium hexafluoride gas..

Mathematical Formula

The general formula for calculating the revolutions per minute (RPM) of a centrifuge is

$$RPM = \sqrt{\frac{g}{r}},$$

where g represents the respective force of the centrifuge and r the radius from the center of the rotor to a point in the sample. However, depending on the centrifuge model used, the respective angle of the rotor and the radius may vary, thus the formula gets modified. For example, the Sorvall #SS-34 rotor has a maximum radius of 10.8 cm, so the formula becomes $RPM = 299\sqrt{\frac{g}{r}}$ which can further simplify to $RPM = 91\sqrt{g}$

Centrifugation in Biological Research

Microcentrifuges

Microcentrifuges are used to process small volumes of biological molecules, cells, or nuclei. Microcentrifuge tubes generally hold 0.5 - 2.0 mL of liquid, and are spun at maximum angular speeds of 12,000–13,000 rpm. Microcentrifuges are small enough to fit on a table-top and have rotors that can quickly change speeds. They may or may not have a refrigeration function.

High-Speed Centrifuges

High-speed or superspeed centrifuges can handle larger sample volumes, from a few tens of millilitres to several litres. Additionally, larger centrifuges can also reach higher angular velocities (around 30,000 rpm). The rotors may come with different adapters to hold various sizes of test tubes, bottles, or microtiter plates.

Fractionation Process

General method of fractionation: Cell sample is stored in a suspension which is:

- Buffered - neutral pH, preventing damage to the structure of proteins including enzymes (which could affect ionic bonds)
- Isotonic (of equal water potential) - this prevents water gain or loss by the organelles
- Cool - reducing the overall activity of enzyme released later in the procedure
  - Cells are homogenised in a blender and filtered to remove debris
  - The homogenised sample is placed in an ultracentrifuge and spun in low speed - nuclei settle out, forming a pellet
  - The supernatant (suspension containing remaining organelles) is spun at a higher speed - chloroplasts settle out
  - The supernatant is spun at a higher speed still - mitochondria and lysosomes settle out
  - The supernatant is spun at an even higher speed - ribosomes, membranes settle out

The ribosomes, membranes and Golgi complexes can be separated by another technique called density gradient centrifugation.

Ultracentrifugations

Ultracentrifugation makes use of high centrifugal force for studying properties of biological particles. Compared to microcentrifuges or high-speed centrifuges, ultracentrifuges can isolate much smaller particles, including ribosomes, proteins, and viruses. Ultracentrifuges can also be used in the study of membrane fractionation. This occurs because ultracentrifuges can reach maximum angular velocities in excess of 70,000 rpm. Additionally, while microcentrifuges and supercentrifuges separate particles in batches (limited volumes of samples must be handled manually in test tubes or bottles), ultracentrifuges can separate molecules in batch or continuous flow systems.

In addition to purification, analytical ultracentrifugation (AUC) can be used for determination of the properties of macromolecules such as shape, mass, composition, and conformation. Samples are centrifuged with a high-density solution such as sucrose, caesium chloride, or iodixanol. The high-density solution may be at a uniform concentration throughout the test tube (cushion) or a varying concentration (gradient). Molecular properties can be modeled through sedimentation velocity analysis or sedimentation equilibrium analysis. During the run, the particle or molecules will migrate through the test tube at different speeds depending on their physical properties and the properties of the solution, and eventually form a pellet at the bottom of the tube, or bands at various heights.

Density Gradient Centrifugation

Density gradient centrifugation Is considered one of the more efficient methods of separating suspended particles. Density gradient centrifugation can be used both as a separation technique and as a method of measuring the densities of particles or molecules in a mixture. A tube, after being centrifuged by this method, has particles in order of density based on height. The object or particle of interest will reside in the position within the tube corresponding to its density.

Linderstorm-Lang, in 1937, discovered that density gradient tubes could be used for density measurements. He discovered this when working with potato yellow-dwarf virus.

This method was also used in Meselson and Stahl's famous experiment in which they proved that DNA replication is semi-conservative by using different isotopes of nitrogen. They used density gradient centrifugation to determine which isotope or isotopes of nitrogen were present in the DNA after cycles of replication.

Nevertheless, some non-ideal sedimentations are still possible when using this method. The first potential issue is the unwanted aggregation of particles, but this can occur in any centrifugation. The second possibility occurs when droplets of solution that contain particles sediment. This is more likely to occur when working with a solution that has a layer of suspension floating on a dense liquid, which in fact have little to no density gradient.

Differential Centrifugation

Differential Centrifugation is a type of centrifugation in which one selectively spins down components of a mixture by a series of increasing centrifugation forces. This method is commonly used to separate organelles and membranes found in cells. Organelles generally differ from each other in density in size, making the use of differential centrifugation, and centrifugation in general, possible. The organelles can then be identified by testing for indicators that are unique to the specific organelles.

Other Applications

- Separating chalk powder from water
- Removing fat from milk to produce skimmed milk
- Separating particles from an air-flow using cyclonic separation
- The clarification and stabilization of wine
- Separation of urine components and blood components in forensic and research laboratories
- Aids in separation of proteins using purification techniques such as salting out, e.g. ammonium sulfate precipitation.

By 1923 Theodor Svedberg and his student H. Rinde had successfully analyzed large-grained sols in terms of their gravitational sedimentation.Sols consist of a substance evenly distributed in another substance, also known as a colloid. However, smaller grained sols, such as those containing gold, could not be analyzed. To investigate this problem Svedberg developed an analytical centrifuge, equipped with a photographic absorption system, which would exert a much greater centrifugal effect. In addition, he developed the theory necessary to measure molecular weight. During this time, Svedberg's attention shifted from gold to proteins.

By 1900, it had been generally accepted that proteins were composed of amino acids; however, whether proteins were colloids or macromolecules was still under debate. One protein being investigated at the time was hemoglobin. It was determined to have 712 carbon, 1,130 hydrogen, 243 oxygen, two sulfur atoms, and at least one iron atom. This gave hemoglobin a resulting weight of approximately 16,000 dalton (Da) but it was uncertain whether this value was a multiple of one or four (dependent upon the number of iron atoms present).

Through a series of experiments utilizing the sedimentation equilibrium technique, two important observations were made: hemoglobin has a molecular weight of 68,000 Da, suggesting that there are four iron atoms present rather than one, and that no matter where the hemoglobin was isolated from, it had exactly the same molecular weight. How something of such a large molecular mass could be consistently found, regardless of where it was sampled from in the body, was unprecedented and favored the idea the proteins are macromolecules rather than colloids. In order to investigate this phenomenon, a centrifuge with even higher speeds was needed, and thus the ultracentrifuge was created to apply the theory of sedimentation-diffusion. The same molecular mass was determined, and the presence of a spreading boundary suggested that it was a single compact particle. Further application of centrifugation showed that under different conditions the large homogeneous particles could be broken down into discrete subunits. The development of centrifugation was a great advance in experimental protein science.

Sources

- Harrison, Roger G., Todd, Paul, Rudge, Scott R., Petrides D.P. *Bioseparations Science and Engineering*. Oxford University Press, 2003.
- Dishon, M., Weiss, G.H., Yphantis, D.A. *Numerical Solutions of the Lamm Equation. I. Numerical Procedure*. Biopolymers, Vol. 4, 1966. pp. 449–455.

- Cao, W., Demeler B. *Modeling Analytical Ultracentrifugation Experiments with an Adaptive Space-Time Finite Element Solution for Multicomponent Reacting Systems*. Biophysical Journal, Vol. 95, 2008. pp. 54–65.
- Cole, J.L., Hansen, J.C. *Analytical Ultracentrifugation as a Contemporary Biomolecular Research Tool*. Methods and Reviews, 1999/2000.
- Howlett, G.J., Minton, A.P., Rivas, G. *Analytical Ultracentrifugation for the Study of Protein Association and Assembly*. Current Opinion in Chemical Biology, Vol. 10, 2006. pp. 430–436.
- Dam, J., Velikovsky, C.A., Mariuzza R.A., et al. *Sedimentation Velocity Analysis of Heterogeneous Protein-Protein Interactions: Lamm Equation Modeling and Sedimentation Coefficient Distributions c(s)*. Biophysical Journal, Vol. 89, 2005. pp. 619–634.
- Berkowitz, S.A., Philo, J.S. *Monitoring the Homogeneity of Adenovirus Preparations (a Gene Therapy Delivery System) Using Analytical Ultracentrifugation*. Analytical Biochemistry, Vol. 362, 2007. pp. 16–37.

CYCLONIC SEPARATION

Cyclonic separation is a method of removing particulates from an air, gas or liquid stream, without the use of filters, through vortex separation. When removing particulate matter from liquids, a hydrocyclone is used; while from gas, a gas cyclone is used. Rotational effects and gravity are used to separate mixtures of solids and fluids. The method can also be used to separate fine droplets of liquid from a gaseous stream.

A high speed rotating (air)flow is established within a cylindrical or conical container called a **cyclone**. Air flows in a helical pattern, beginning at the top (wide end) of the cyclone and ending at the bottom (narrow) end before exiting the cyclone in a straight stream through the center of the cyclone and out the top. Larger (denser) particles in the rotating stream have too much inertia to follow the tight curve of the stream, and strike the outside wall, then fall to the bottom of the cyclone where they can be removed. In a conical system, as the rotating flow moves towards the narrow end of the cyclone, the rotational radius of the stream is reduced, thus separating smaller and smaller particles. The cyclone geometry, together with flow rate, defines the *cut point* of the cyclone. This is the size of particle that will be removed from the stream with a 50% efficiency. Particles larger than the cut point will be removed with a greater efficiency, and smaller particles with a lower efficiency.

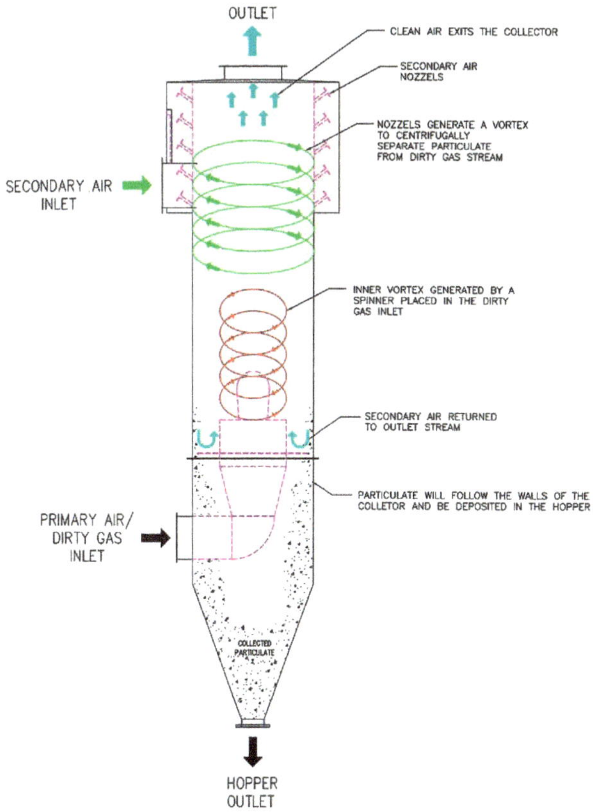

Airflow diagram for Aerodyne cyclone in standard vertical position. Secondary air flow is injected to reduce wall abrasion.

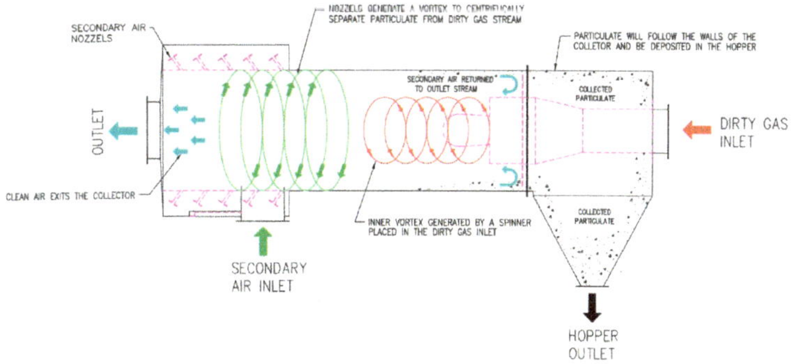

Airflow diagram for Aerodyne cyclone in horizontal position, an alternate design. Secondary air flow is injected to reduce wall abrasion, and to help move collected particulates to hopper for extraction.

An alternative cyclone design uses a secondary air flow within the cyclone to keep the collected particles from striking the walls, to protect them from abrasion. The primary air flow containing the particulates enters from the bottom of the cyclone and is forced into spiral rotation by stationary spinner vanes. The secondary air flow enters from the top of the cyclone and moves downward toward the bottom, intercepting the particulate from the primary air. The secondary air flow also allows the collector to optionally be mounted horizontally, because it pushes the particulate toward the collection area, and does not rely solely on gravity to perform this function.

Large scale cyclones are used in sawmills to remove sawdust from extracted air. Cyclones are also used in oil refineries to separate oils and gases, and in the cement industry as components of kiln preheaters. Cyclones are increasingly used in the household, as the core technology in bagless types of portable vacuum cleaners and central vacuum cleaners. Cyclones are also used in industrial and professional kitchen ventilation for separating the grease from the exhaust air in extraction hoods. Smaller cyclones are used to separate airborne particles for analysis. Some are small enough to be worn clipped to clothing, and are used to separate respirable particles for later analysis.

Similar separators are used in the oil refining industry (e.g. for Fluid catalytic cracking) to achieve fast separation of the catalyst particles from the reacting gases and vapors.

James Dyson has become a billionaire from developing and marketing bagless vacuum cleaners based on cyclonic separation of dust, initially inspired by seeing sawdust separator at a sawmill.

Analogous devices for separating particles or solids from liquids are called hydrocyclones or hydroclones. These may be used to separate solid waste from water in wastewater and sewage treatment.

Cyclone Theory

As the cyclone is essentially a two phase particle-fluid system, fluid mechanics and particle transport equations can be used to describe the behaviour of a cyclone. The air in a cyclone is initially introduced tangentially into the cyclone with an inlet velocity V_{in}. Assuming that the particle is spherical, a simple analysis to calculate critical separation particle sizes can be established.

If one considers an isolated particle circling in the upper cylindrical component of the cyclone at a rotational radius of r from the cyclone's central axis, the particle is therefore subjected to drag, centrifugal, and buoyant forces. Given that the fluid velocity is moving in a spiral the gas velocity can be broken into two component velocities: a tangential component, V_t, and an outward radial velocity component V_r. Assuming Stokes' law, the drag force in the outward radial direction that is opposing the outward velocity on any particle in the inlet stream is:

$$F_d = -6\pi r_p \mu V_r.$$

Using ρ_p as the particles density, the centrifugal component in the outward radial direction is:

$$\begin{aligned} F_c &= m\frac{V_t^2}{r} \\ &= \frac{4}{3}\pi\rho_p r_p^3 \frac{V_t^2}{r}. \end{aligned}$$

The buoyant force component is in the inward radial direction. It is in the opposite direction to the particle's centrifugal force because it is on a volume of fluid that is missing compared to the surrounding fluid. Using ρ_f for the density of the fluid, the buoyant force is:

$$F_b = -V_p \rho_f \frac{V_t^2}{r}$$

$$= -\frac{4\pi r_p^3}{3} \frac{V_t^2}{r} \rho_f .$$

In this case, V_p is equal to the volume of the particle (as opposed to the velocity). Determining the outward radial motion of each particle is found by setting Newton's second law of motion equal to the sum of these forces:

$$m\frac{dV_r}{dt} = F_d + F_c + F_b$$

To simplify this, we can assume the particle under consideration has reached "terminal velocity", i.e., that its acceleration $\frac{dV_r}{dt}$ is zero. This occurs when the radial velocity has caused enough drag force to counter the centrifugal and buoyancy forces. This simplification changes our equation to:

$$F_d + F_c + F_b = 0$$

Which expands to:

$$-6\pi r_p \mu V_r + \frac{4}{3}\pi r_p^3 \frac{V_t^2}{r}\rho_p - \frac{4}{3}\pi r_p^3 \frac{V_t^2}{r}\rho_f = 0$$

Solving for V_r we have

$$V_r = \frac{2}{9}\frac{r_p^2}{\mu}\frac{V_t^2}{r}(\rho_p - \rho_f).$$

Notice that if the density of the fluid is greater than the density of the particle, the motion is (-), toward the center of rotation and if the particle is denser than the fluid, the motion is (+), away from the center. In most cases, this solution is used as guidance in designing a separator, while actual performance is evaluated and modified empirically.

In non-equilibrium conditions when radial acceleration is not zero, the general equation from above must be solved. Rearranging terms we obtain

$$\frac{dV_r}{dt} + \frac{9}{2}\frac{\mu}{\rho_p r_p^2}V_r - \left(1 - \frac{\rho_f}{\rho_p}\right)\frac{V_t^2}{r} = 0$$

Since V_r is distance per time, this is a 2nd order differential equation of the form $x'' + c_1 x' + c_2 = 0$

Experimentally it is found that the velocity component of rotational flow is proportional to r_2, therefore:

$$V_t \propto r^2.$$

This means that the established feed velocity controls the vortex rate inside the cyclone, and the velocity at an arbitrary radius is therefore:

$$U_r = U_{in} \frac{r}{R_{in}}.$$

Subsequently, given a value for V_t, possibly based upon the injection angle, and a cutoff radius, a characteristic particle filtering radius can be estimated, above which particles will be removed from the gas stream.

Alternative Models

The above equations are limited in many regards. For example, the geometry of the separator is not considered, the particles are assumed to achieve a steady state and the effect of the vortex inversion at the base of the cyclone is also ignored, all behaviours which are unlikely to be achieved in a cyclone at real operating conditions.

More complete models exist, as many authors have studied the behaviour of cyclone separators. Numerical modelling using computational fluid dynamics has also been used extensively in the study of cyclonic behaviour. A major limitation of any fluid mechanics model for cyclone separators is the inability to predict the agglomeration of fine particles with larger particles, which has a great impact on cyclone collection efficiency.

CHROMATOGRAPHY

Chromatography is a laboratory technique for the separation of a mixture. The mixture is dissolved in a fluid called the *mobile phase,* which carries it through a structure holding another material called the *stationary phase.* The various constituents of the mixture travel at different speeds, causing them to separate. The separation is based on differential partitioning between the mobile and stationary phases. Subtle differences in a compound's partition coefficient result in differential retention on the stationary phase and thus changing the separation.

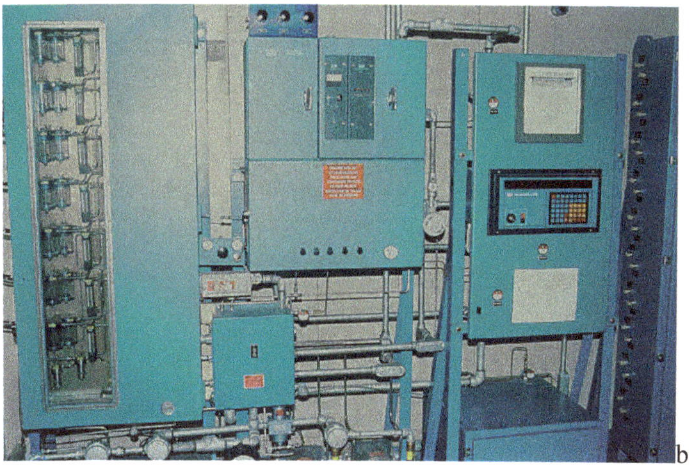

b

Pictured is a sophisticated gas chromatography system. This instrument records concentrations of acrylonitrile in the air at various points throughout the chemical laboratory.

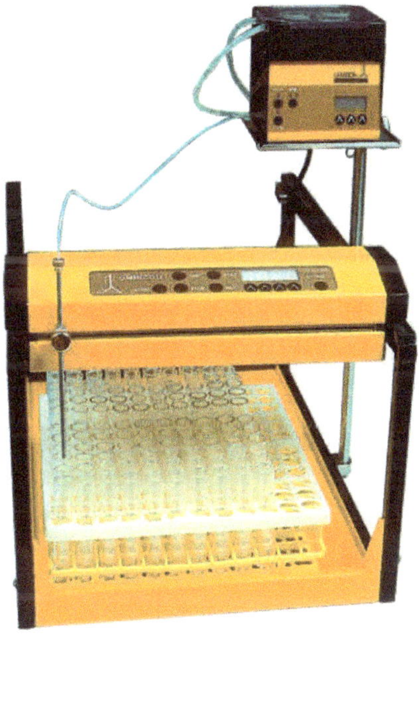

Automated fraction collector and sampler for chromatographic techniques.

Chromatography may be preparative or analytical. The purpose of preparative chromatography is to separate the components of a mixture for more advanced use (and is thus a form of purification). Analytical chromatography is done normally with smaller amounts of material and is for measuring the relative proportions of analytes in a mixture. The two are not mutually exclusive.

HISTORY

Thin layer chromatography is used to separate components of a plant extract, illustrating the experiment with plant pigments that gave chromatography its name.

Chromatography was first employed in Russia by the Italian-born scientist Mikhail Tsvet in 1900. He continued to work with chromatography in the first decade of the 20th century, primarily for the separation of plant pigments such as chlorophyll, carotenes, and xanthophylls. Since these components have different colors (green, orange, and yellow, respectively) they gave the technique its name. New types of chromatography developed during the 1930s and 1940s made the technique useful for many separation processes.

Chromatography technique developed substantially as a result of the work of Archer John Porter Martin and Richard Laurence Millington Synge during the 1940s and 1950s, for which they won a Nobel prize. They established the principles and basic techniques of partition chromatography, and their work encouraged the rapid development of several chromatographic methods: paper chromatography, gas chromatography, and what would become known as high performance liquid chromatography. Since then, the technology has advanced rapidly. Researchers found that the main principles of Tsvet›s chromatography could be applied in many different ways, resulting in the different varieties of chromatography described below. Advances are continually improving the technical performance of chromatography, allowing the separation of increasingly similar molecules.

CHROMATOGRAPHY TERMS

- The **analyte** is the substance to be separated during chromatography. It is also normally what is needed from the mixture.
- **Analytical chromatography** is used to determine the existence and possibly also the concentration of analyte(s) in a sample.

- A **bonded phase** is a stationary phase that is covalently bonded to the support particles or to the inside wall of the column tubing.
- A **chromatogram** is the visual output of the chromatograph. In the case of an optimal separation, different peaks or patterns on the chromatogram correspond to different components of the separated mixture.

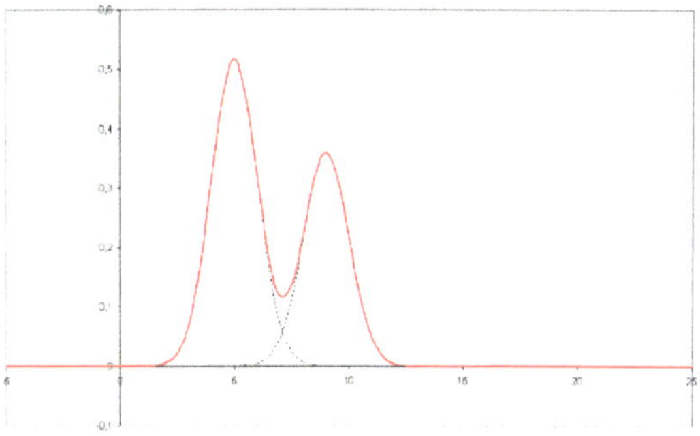

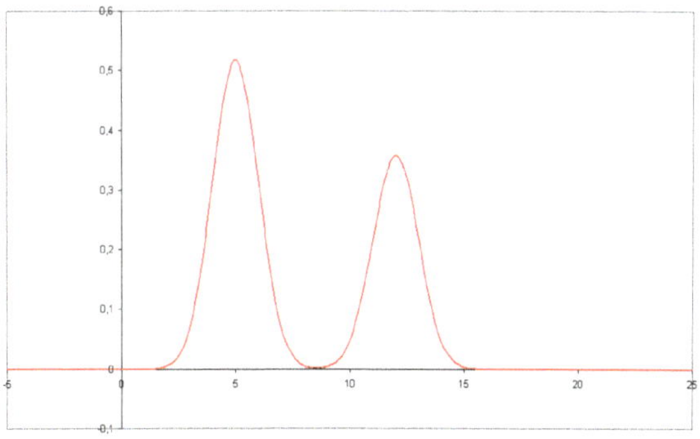

Plotted on the x-axis is the retention time and plotted on the y-axis a signal (for example obtained by a spectrophotometer, mass spectrometer or a variety of other detectors) corresponding to the response created by the analytes exiting the system. In the case of an optimal system the signal is proportional to the concentration of the specific analyte separated.

- A **chromatograph** is equipment that enables a sophisticated separation, e.g. gas chromatographic or liquid chromatographic separation.
- **Chromatography** is a physical method of separation that distributes components to separate between two phases, one stationary (stationary phase), the other (the mobile phase) moving in a definite direction.
- The **eluate** is the mobile phase leaving the column.
- The **eluent** is the solvent that carries the analyte.
- An **eluotropic series** is a list of solvents ranked according to their eluting power.
- An **immobilized phase** is a stationary phase that is immobilized on the support particles, or on the inner wall of the column tubing.
- The **mobile phase** is the phase that moves in a definite direction. It may be a liquid (LC and Capillary Electrochromatography (CEC)), a gas (GC), or a supercritical

fluid (supercritical-fluid chromatography, SFC). The mobile phase consists of the sample being separated/analyzed and the solvent that moves the sample through the column. In the case of HPLC the mobile phase consists of a non-polar solvent(s) such as hexane in normal phase or a polar solvent such as methanol in reverse phase chromatography and the sample being separated. The mobile phase moves through the chromatography column (the stationary phase) where the sample interacts with the stationary phase and is separated.

- **Preparative chromatography** is used to purify sufficient quantities of a substance for further use, rather than analysis.
- The **retention time** is the characteristic time it takes for a particular analyte to pass through the system (from the column inlet to the detector) under set conditions: Kovats› retention index
- The **sample** is the matter analyzed in chromatography. It may consist of a single component or it may be a mixture of components. When the sample is treated in the course of an analysis, the phase or the phases containing the analytes of interest is/are referred to as the sample whereas everything out of interest separated from the sample before or in the course of the analysis is referred to as waste.
- The **solute** refers to the sample components in partition chromatography.
- The **solvent** refers to any substance capable of solubilizing another substance, and especially the liquid mobile phase in liquid chromatography.
- The **stationary phase** is the substance fixed in place for the chromatography procedure. Examples include the silica layer in thin layer chromatography
- The **detector** refers to the instrument used for qualitative and quantitative detection of analytes after separation.

Chromatography is based on the concept of partition coefficient. Any solute partitions between two immiscible solvents. When we make one solvent immobile (by adsorption on a solid support matrix) and another mobile it results in most common applications of chromatography. If the matrix support, or stationary phase, is polar (e.g. paper, silica etc.) it is forward phase chromatography, and if it is non-polar (C-18) it is reverse phase.

DISPLACEMENT CHROMATOGRAPHY

The basic principle of displacement chromatography is: A molecule with a high affinity for the chromatography matrix (the displacer) competes effectively for binding sites, and thus displace all molecules with lesser affinities. There are distinct differences between displacement and elution chromatography. In elution mode, substances typically emerge from a column in narrow, Gaussian peaks. Wide separation of peaks, preferably to baseline, is desired for maximum purification. The speed at which any component of a mixture travels down the column in elution mode depends on many factors. But for two substances to travel at different speeds, and thereby be resolved, there must be substantial differences in some interaction between the biomolecules and the chromatography matrix. Operating parameters are adjusted to maximize the effect of this difference. In many cases, baseline separation of the peaks can be achieved only with gradient elution and low column loadings. Thus, two drawbacks to elution mode chromatography, especially at the preparative scale, are operational complexity, due to gradient solvent pumping, and low throughput, due to low column loadings. Displacement chromatography has advantages over elution chromatography in that components are resolved into consecutive zones of pure substances rather than "peaks". Because the process takes advantage of the nonlinearity of the isotherms, a larger column feed can be separated on a given column with the purified components recovered at significantly higher concentrations.

TECHNIQUES BY SEPARATION MECHANISM

Ion Exchange Chromatography

Ion exchange chromatography (usually referred to as ion chromatography) uses an ion exchange mechanism to separate analytes based on their respective charges. It is usually performed in columns but can also be useful in planar mode. Ion exchange chromatography uses a charged stationary phase to separate charged compounds including anions, cations, amino acids, peptides, and proteins. In conventional methods the stationary phase is an ion exchange resin that carries charged functional groups that interact with oppositely charged groups of the compound to retain. Ion exchange chromatography is commonly used to purify proteins using FPLC.

Size-Exclusion Chromatography

Size-exclusion chromatography (SEC) is also known as gel permeation chromatography (GPC) or **gel filtration chromatography** and separates molecules according to their size (or more accurately according to their hydrodynamic diameter or hydrodynamic volume). Smaller molecules are able to enter the pores of the media and, therefore, molecules are trapped and removed from the flow of the mobile phase. The average residence time in the pores depends upon the effective size of the analyte molecules. However, molecules that are larger than the average pore size of the packing are excluded and thus suffer essentially no retention; such species are the first to be eluted. It is generally a low-resolution chromatography technique and thus it is often reserved for the final, polishing step of a purification. It is also useful for determining the tertiary structure and quaternary structure of purified proteins, especially since it can be carried out under native solution conditions.

Expanded Bed Adsorption Chromatographic Separation

An expanded bed chromatographic adsorption (EBA) column for a biochemical separation process comprises a pressure equalization liquid distributor having a self-cleaning function below a porous blocking sieve plate at the bottom of the expanded bed, an upper part nozzle assembly having a backflush cleaning function at the top of the expanded bed, a better distribution of the feedstock liquor added into the expanded bed ensuring that the fluid passed through the expanded bed layer displays a state of piston flow. The expanded bed layer displays a state of piston flow. The expanded bed chromatographic separation column has advantages of increasing the separation efficiency of the expanded bed.

Expanded-bed adsorption (EBA) chromatography is a convenient and effective technique for the capture of proteins directly from unclarified crude sample. In EBA chromatography, the settled bed is first expanded by upward flow of equilibration buffer. The crude feed, a mixture of soluble proteins, contaminants, cells, and cell debris, is then passed upward through the expanded bed. Target proteins are captured on the adsorbent, while particulates and contaminants pass through. A change to elution buffer while maintaining upward flow results in desorption of the target protein in expanded-bed mode. Alternatively, if the flow is reversed, the adsorbed particles will quickly settle and the proteins can be desorbed by an elution buffer. The mode used for elution (expanded-bed versus settled-bed) depends on the characteristics of the feed. After elution, the adsorbent is cleaned with a predefined cleaning-in-place (CIP) solution, with cleaning followed by either column regeneration or storage.

Special Techniques

Reversed-Phase Chromatography

Reversed-phase chromatography (RPC) is any liquid chromatography procedure in which the mobile phase is significantly more polar than the stationary phase. It is so named because in normal-phase liquid chromatography, the mobile phase is significantly less polar than the stationary phase. Hydrophobic molecules in the mobile phase tend to adsorb to the relatively hydrophobic stationary phase. Hydrophilic molecules in the mobile phase will tend to elute first. Separating columns typically comprise a C8 or C18 carbon-chain bonded to a silica particle substrate.

Hydrophobic Interaction Chromatography

Hydrophobic interactions between proteins and the chromatographic matrix can be exploited to purify proteins. In hydrophobic interaction chromatography the matrix material is lightly substituted with hydrophobic groups. These groups can range from methyl, ethyl, propyl, octyl, or phenyl groups. At high salt concentrations, non-polar sidechains on the surface on proteins "interact" with the hydrophobic groups; that is, both types of groups are excluded by the polar solvent (hydrophobic effects are augmented by increased ionic strength). Thus, the sample is applied to the column in a buffer which is highly polar. The eluant is typically an aqueous buffer with decreasing salt concentrations, increasing concentrations of detergent (which disrupts hydrophobic interactions), or changes in pH.

In general, Hydrophobic Interaction Chromatography (HIC) is advantageous if the sample is sensitive to pH change or harsh solvents typically used in other types of chromatography but not high salt concentrations. Commonly, it is the amount of salt in the buffer which is varied. In 2012, Müller and Franzreb described the effects of temperature on HIC using Bovine Serum Albumin (BSA) with four different types of hydrophobic resin. The study altered temperature as to effect the binding affinity of BSA onto the matrix. It was concluded that cycling temperature from 50 degrees to 10 degrees would not be adequate to effectively wash all BSA from the matrix but could be very effective if the column would only be used a few times. Using temperature to effect change allows labs to cut costs on buying salt and saves money.

If high salt concentrations along with temperature fluctuations want to be avoided you can use a more hydrophobic to compete with your sample to elute it. source This so-called salt independent method of HIC showed a direct isolation of Human Immunoglobulin G (IgG) from serum with satisfactory yield and used Beta-cyclodextrin as a competitor to displace IgG from the matrix. This largely opens up the possibility of using HIC with samples which are salt sensitive as we know high salt concentrations precipitate proteins.

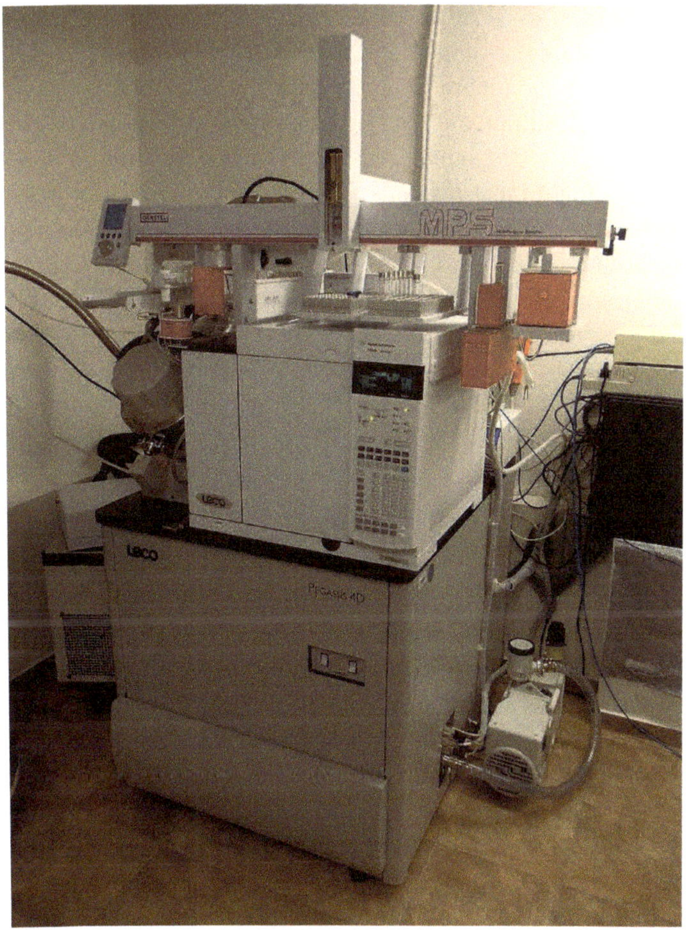

Two-dimensional chromatograph GCxGC-TOFMS at Chemical Faculty of GUT Gdańsk, Poland, 2016

Two-Dimensional Chromatography

In some cases, the chemistry within a given column can be insufficient to separate some analytes. It is possible to direct a series of unresolved peaks onto a second column with different physico-chemical (Chemical classification) properties. Since the mechanism of retention on this new solid support is different from the first dimensional separation, it can be possible to separate compounds that are indistinguishable by one-dimensional chromatography. The sample is spotted at one corner of a square plate,developed, air-dried, then rotated by 90° and usually redeveloped in a second solvent system.

Simulated Moving-Bed Chromatography

The simulated moving bed (SMB) technique is a variant of high performance liquid chromatography; it is used to separate particles and/or chemical compounds that would be difficult or impossible to resolve otherwise. This increased separation is brought about by a valve-and-column arrangement that is used to lengthen the stationary phase indefinitely. In the moving bed technique of preparative chromatography the feed entry and the analyte recovery are simultaneous and continuous, but because of practical difficulties with a continuously moving bed, simulated moving bed technique was proposed. In the simulated moving bed technique instead of moving the bed, the sample inlet and the analyte exit positions are moved continuously, giving the impression of a moving bed. True moving bed chromatography (TMBC) is only a theoretical concept. Its simulation, SMBC is achieved by the use of a multiplicity of columns in series and a complex valve arrangement, which provides for sample and solvent feed, and also analyte and waste takeoff at appropriate locations of any column, whereby it allows switching at regular intervals the sample entry in one direction, the solvent entry in the opposite direction, whilst changing the analyte and waste takeoff positions appropriately as well.

Pyrolysis Gas Chromatography

Pyrolysis gas chromatography mass spectrometry is a method of chemical analysis in which the sample is heated to decomposition to produce smaller molecules that are separated by gas chromatography and detected using mass spectrometry.

Pyrolysis is the thermal decomposition of materials in an inert atmosphere or a vacuum. The sample is put into direct contact with a platinum wire, or placed in a quartz sample tube, and rapidly heated to 600–1000 °C. Depending on the application even higher temperatures are used. Three different heating techniques are used in actual pyrolyzers: Isothermal furnace, inductive heating (Curie Point filament), and resistive heating using platinum filaments. Large molecules cleave at their weakest points and produce smaller, more volatile fragments. These fragments can be separated by gas chromatography. Pyrolysis GC chromatograms are typically complex because a wide range of different decomposition products is formed. The data can either be used as fingerprint to prove material identity or the GC/MS data is used to identify individual fragments to obtain structural information. To increase the volatility of polar fragments, various methylating reagents can be added to a sample before pyrolysis.

Besides the usage of dedicated pyrolyzers, pyrolysis GC of solid and liquid samples can be performed directly inside Programmable Temperature Vaporizer (PTV) injectors that provide quick heating (up to 30 °C/s) and high maximum temperatures of 600–650 °C. This is sufficient for some pyrolysis applications. The main advantage is that no dedicated instrument has to be purchased and pyrolysis can be performed as part of routine GC analysis. In this case quartz GC inlet liners have to be used. Quantitative data can be acquired, and good results of derivatization inside the PTV injector are published as well.

Fast Protein Liquid Chromatography

Fast protein liquid chromatography (FPLC), is a form of liquid chromatography that is often used to analyze or purify mixtures of proteins. As in other forms of chromatography, separation is possible because the different components of a mixture have different affinities for two materials, a moving fluid (the "mobile phase") and a porous solid (the stationary phase). In FPLC the mobile phase is an aqueous solution, or "buffer". The buffer flow rate is controlled by a positive-displacement pump and is normally kept constant, while the composition of the buffer can be varied by drawing fluids in different proportions from two or more external reservoirs. The stationary phase is a resin composed of beads, usually of cross-linked agarose, packed into a cylindrical glass or plastic column. FPLC resins are available in a wide range of bead sizes and surface ligands depending on the application.

Countercurrent Chromatography

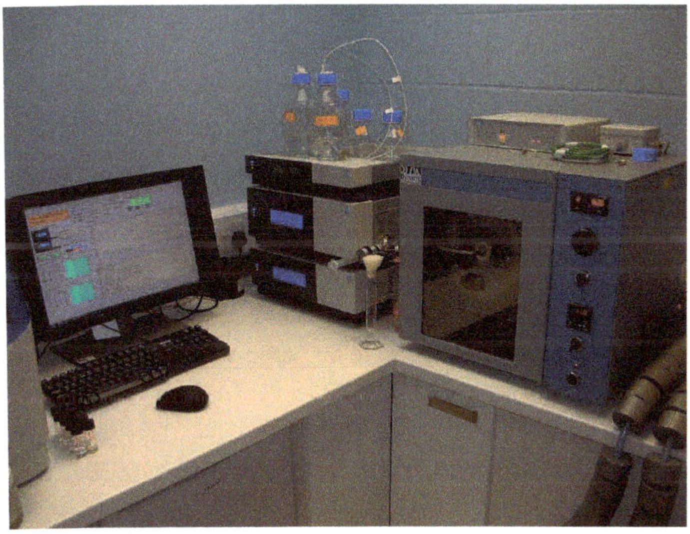

An example of a HPCCC system

Countercurrent chromatography (CCC) is a type of liquid-liquid chromatography, where both the stationary and mobile phases are liquids. The operating principle of CCC equipment requires a column consisting of an open tube coiled around a bobbin. The bobbin is rotated in a double-axis gyratory motion (a cardioid), which causes a variable gravity (G) field to act on the column during each rotation. This motion causes the column to see one partitioning step per revolution and components of the sample separate in the column due to their partitioning coefficient between the two immiscible liquid phases used. There are many types of CCC available today. These include HSCCC (High Speed CCC) and HPCCC (High Performance CCC). HPCCC is the latest and best performing version of the instrumentation available currently.

Chiral Chromatography

Chiral chromatography involves the separation of stereoisomers. In the case of enantiomers, these have no chemical or physical differences apart from being three-dimensional mirror images. Conventional chromatography or other separation processes are incapable of separating them. To enable chiral separations to take place, either the mobile phase or the stationary phase must themselves be made chiral, giving differing affinities between the analytes. Chiral chromatography HPLC columns (with a chiral stationary phase) in both normal and reversed phase are commercially available.

REFERENCES

1. Garrett, Reginald H.; Grisham, Charles M. (2013). Biochemistry (5th ed.). Belmont, CA: Brooks/Cole, Cengage Learning. p. 111. ISBN 9781133106296.
2. Frei, Mark. "Centrifugation Basics". Sigma-Aldrich. Retrieved 10 May 2016.
3. Article on "Centrifugation" retrieved on 15th October 2013 from http://www.lenntech.com/library/clarification/clarification/centrifugation.htm
4. Ballou, David P.; Benore, Marilee; Ninfa, Alexander J. (2008). Fundamental laboratory approaches for biochemistry and biotechnology. (2nd ed.). Hoboken, N.J.: Wiley. p. 43. ISBN 9780470087664.
5. Ballou, David P.; Benore, Marilee; Ninfa, Alexander J. (2008). Fundamental laboratory approaches for biochemistry and biotechnology. (2nd ed.). Hoboken, N.J.: Wiley. p. 235. ISBN 9780470087664.
6. Brakke, Myron K. (April 1951). "Density Gradient Centrifugation: A New Separation Technique". J. Am. Che. Soc. 73 (4): 1847–1848. doi:10.1021/ja01148a508.
7. Oster, Gerald; Yamamoto, Masahide (June 1963). "Density Gradient Techniques". Chem. Rev. 63 (3): 257–268. doi:10.1021/cr60223a003.
8. Ballou, David P.; Benore, Marilee; Ninfa, Alexander J. (2010). Fundamental Laboratory Approaches for Biochemistry and Biotechnology (seconde ed.). University of Michigan: John Wiley & Sons, Inc. p. 213. ISBN 978-0-470-08766-4.
9. Ballou, David P.; Benore, Marilee; Ninfa, Alexander J. (2008). Fundamental laboratory approaches for biochemistry and biotechnology. (2nd ed.). Hoboken, N.J.: Wiley. pp. 238–239. ISBN 9780470087664.
10. Van Holde, K. E. (1998). Analytical ultracentrifugation from 1924 to the present: A remarkable history. Chemtracts – Biochemistry and Molecular Biology. 11:933-943
11. Svedberg, T. (1927). The Ultracentrifuge Nobel Lecture

12. Tanford, C., and Reynolds, J. 2001. Nature's robots: A history of proteins. Oxford University Press. pp. 303-305
13. Simoni, D. S., Hill, R. L., and Vaughan, M. (2002). The structure and function of hemoglobin: Gilbery Smithson Adair and the Adair equations. The Journal of Biological Chemistry. 277(31): e1-e2
14. "chromatography". Online Etymology Dictionary.
15. McMurry, John (2011). Organic chemistry: with biological applications (2nd ed.). Belmont, CA: Brooks/Cole. p. 395. ISBN 9780495391470.
16. Hostettmann, K; Marston, A; Hostettmann, M (1998). Preparative Chromatography Techniques Applications in Natural Product Isolation (Second ed.). Berlin, Heidelberg: Springer Berlin Heidelberg. p. 50. ISBN 9783662036310.
17. The first report of Tsvet's discovery was published in 1905:
18. Ettre, L. S.; Sakodynskii, K. I. (March 1993). "M. S. Tswett and the discovery of chromatography II: Completion of the development of chromatography (1903–1910)". Chromatographia. 35 (5-6): 329–338. doi:10.1007/BF02277520.
19. "The Nobel Prize in Chemistry 1952". nobelprize.org. Retrieved 25 August 2016.
20. Ettre, L. S. (1993). "Nomenclature for chromatography (IUPAC Recommendations 1993)". Pure and Applied Chemistry. 65 (4). doi:10.1351/pac199365040819.
21. Still, W. C.; Kahn, M.; Mitra, A. (1978). "Rapid chromatographic technique for preparative separations with moderate resolution". J. Org. Chem. 43 (14): 2923–2925. doi:10.1021/jo00408a041.
22. Harwood, Laurence M.; Moody, Christopher J. (1989). Experimental organic chemistry: Principles and Practice (Illustrated ed.). WileyBlackwell. pp. 180–185. ISBN 978-0-632-02017-1.
23. Anfinsen, Christian B.; Edsall, John Tileston; Richards, Frederic Middlebrook, eds. (1976). Advances in Protein Chemistry. pp. 6–7. ISBN 978-0-12-034230-3.
24. Displacement Chromatography 101 Archived 15 September 2008 at the Wayback Machine.. Sachem, Inc. Austin, TX 78737
25. Bailon, Pascal; Ehrlich, George K.; Fung, Wen-Jian and Berthold, Wolfgang (2000) An Overview of Affinity Chromatography, Humana Press. ISBN 978-0-89603-694-9.
26. Ninfa, Alexander J., David P. Ballou, and Marilee Benore. Fundamental Laboratory Approaches for Biochemistry and Biotechnology. Hoboken, NJ: John Wiley, 2010. Print.
27. Müller, Tobias K.h., and Matthias Franzreb. "Suitability of Commercial Hydrophobic Interaction Sorbents for Temperature-controlled Protein Liquid Chromatography under Low Salt Conditions." Journal of Chromatography A 1260 (2012): 88-96. Web. <http://ac.els-cdn.com/S002196731201268X/1-s2.0-S002196731201268X-main.pdf?_tid=2f7e7186-1c7e-11e6-836f-00000aacb361&acdnat=1463524077_be16e37f7efe44d6b689342fa7669726>.

Crystallization Process

Chapter Contents

Crystallization refers to the formation of solid crystals from a homogeneous solution. It is essentially a solid-liquid separation technique and a very important one at that. Crystals are grown in many shapes, which are dependent upon downstream processing or final product requirements. Crystal shapes can include cubic, tetragonal, orthorhombic, hexagonal, monoclinic, triclinic, and trigonal. In order for crystallization to take place a solution must be "supersaturated". Supersaturation refers to a state in which the liquid (solvent) contains more dissolved solids (solute) than can ordinarily be accomodated at that temperature.

INTRODUCTION

Crystallization is the (natural or artificial) process where a solid forms where the atoms or molecules are highly organized in a structure known as a crystal. Some of the ways which crystals form are through precipitating from a solution, melting or more rarely deposition directly from a gas. Crystallization is also a chemical solid–liquid separation technique, in which mass transfer of a solute from the liquid solution to a pure solid crystalline phase occurs. In chemical engineering crystallization occurs in a crystallizer. Crystallization is therefore related to precipitation, although the result is not amorphous or disordered, but a crystal.

PROCESS

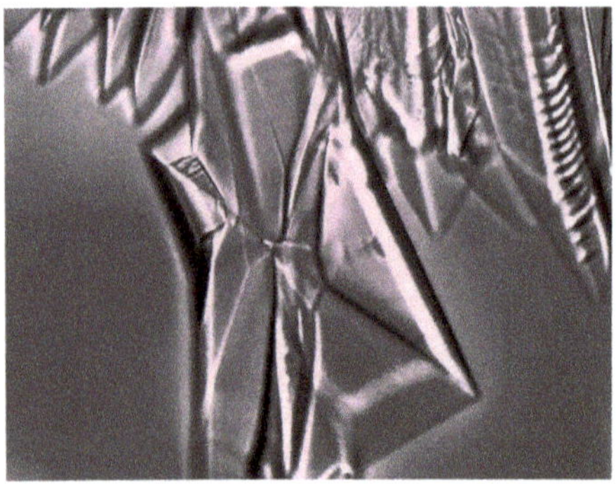

Time-lapse of growth of a citric acid crystal. The video covers an area of 2.0 by 1.5 mm and was captured over 7.2 min.

The crystallization process consists of two major events, *nucleation* and *crystal growth* which are driven by thermodynamic properties as well as chemical properties. In crystallization *Nucleation* is the step where the solute molecules or atoms dispersed in the solvent start to gather into clusters, on the microscopic scale (elevating solute concentration in a small region), that become stable under the current operating conditions. These stable clusters constitute the nuclei. Therefore, the clusters need to reach a critical size in order to become stable nuclei. Such critical size is dictated by many different factors (temperature, supersaturation, etc.). It is at the stage of nucleation that the atoms or molecules arrange in a defined and periodic manner that defines the crystal structure — note that "crystal structure" is a special term that refers to the relative arrangement of the atoms or molecules, not the macroscopic properties of the crystal (size and shape), although those are a result of the internal crystal structure.

The *crystal growth* is the subsequent size increase of the nuclei that succeed in achieving the critical cluster size. Crystal growth is a dynamic process occurring in equilibrium where solute molecules or atoms precipitate out of solution, and dissolve back into solution. Supersaturation is one of the driving forces of crystallization, as the solubility of a species is an equilibrium process quantified by Ksp. Depending upon the conditions, either nucleation or growth may be predominant over the other, dictating crystal size.

Many compounds have the ability to crystallize with some having different crystal structures, a phenomenon called polymorphism. Each polymorph is in fact a different thermodynamic solid state and crystal polymorphs of the same compound exhibit different physical properties, such as dissolution rate, shape (angles between facets and facet growth rates), melting point, etc. For this reason, polymorphism is of major importance in industrial manufacture of crystalline products. Additionally, crystal phases can sometimes be interconverted by varying factors such as temperature.

In Nature

Snowflakes are a very well-known example, where subtle differences in *crystal growth* conditions result in different geometries.

Crystallized honey

There are many examples of natural process that involve crystallization.

Geological time scale process examples include:

- Natural (mineral) crystal formation (gemstone);
- Stalactite/stalagmite, rings formation.

Usual time scale process examples include:

- Snow flakes formation;
- Honey crystallization (nearly all types of honey crystallize).

Methods

Crystal formation can be divided into two types, where the first type of crystals are composed of a cation and anion, also known as a salt, such as sodium acetate. The second type of crystals are composed of uncharged species, for example menthol.

Crystal formation can be achieved by various methods, such as: cooling, evaporation, addition of a second solvent to reduce the solubility of the solute (technique known as antisolvent or drown-out), solvent layering, sublimation, changing the cation or anion, as well as other methods.

The formation of a supersaturated solution does not guarantee crystal formation, and often a seed crystal or scratching the glass is required to form nucleation sites.

A typical laboratory technique for crystal formation is to dissolve the solid in a solution in which it is partially soluble, usually at high temperatures to obtain supersaturation. The hot mixture is then filtered to remove any insoluble impurities. The filtrate is allowed to slowly cool. Crystals that form are then filtered and washed with a solvent in which they are not soluble, but is miscible with the mother liquor. The process is then repeated to increase the purity in a technique known as recrystallization.

Typical Equipment

Equipment for the main industrial processes for crystallization.

- *Tank crystallizers.* Tank crystallization is an old method still used in some specialized cases. Saturated solutions, in tank crystallization, are allowed to cool in open tanks. After a period of time the mother liquor is drained and the crystals removed. Nucleation and size of crystals are difficult to control. Typically, labor costs are very high.

Thermodynamic View

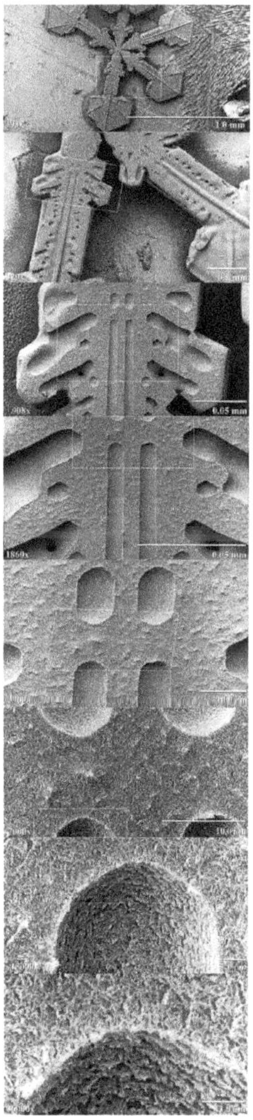

Low-temperature SEM magnification series for a snow crystal. The crystals are captured, stored, and sputter coated with platinum at cryo-temperatures for imaging.

The nature of a crystallization process is governed by both thermodynamic and kinetic factors, which can make it highly variable and difficult to control. Factors such as impurity level, mixing regime, vessel design, and cooling profile can have a major impact on the size, number, and shape of crystals produced.

Now put yourself in the place of a molecule within a pure and *perfect crystal*, being heated by an external source. At some sharply defined temperature, the complicated architecture of the crystal collapses to that of a liquid. Textbook thermodynamics says that melting occurs because the entropy, S, gain in your system by spatial randomization of the molecules has overcome the enthalpy, H, loss due to breaking the crystal packing forces:

$$T(S_{liquid} - S_{solid}) > H_{liquid} - H_{solid}$$

$$G_{liquid} < G_{solid}$$

This rule suffers no exceptions when the temperature is rising. By the same token, on cooling the melt, at the very same temperature the bell should ring again, and molecules should click back into the very same crystalline form. The entropy decrease due to the ordering of molecules within the system is overcompensated by the thermal randomization of the surroundings, due to the release of the heat of fusion; the entropy of the universe increases.

But liquids that behave in this way on cooling are the exception rather than the rule; in spite of the second principle of thermodynamics, crystallization usually occurs at lower temperatures (supercooling). This can only mean that a crystal is more easily destroyed than it is formed. Similarly, it is usually much easier to dissolve a perfect crystal in a solvent than to grow again a good crystal from the resulting solution. The nucleation and growth of a crystal are under kinetic, rather than thermodynamic, control.

Dynamics

As mentioned above, a crystal is formed following a well-defined pattern, or structure, dictated by forces acting at the molecular level. As a consequence, during its formation process the crystal is in an environment where the solute concentration reaches a certain critical value, before changing status. Solid formation, impossible below the solubility threshold at the given temperature and pressure conditions, may then take place at a concentration higher than the theoretical solubility level. The difference between the actual value of the solute concentration at the crystallization limit and the theoretical (static) solubility threshold is called supersaturation and is a fundamental factor in crystallization.

Nucleation

Nucleation is the initiation of a phase change in a small region, such as the formation of a solid crystal from a liquid solution. It is a consequence of rapid local fluctuations on a molecular scale in a homogeneous phase that is in a state of metastable equilibrium. Total nucleation is the sum effect of two categories of nucleation – primary and secondary.

Primary Nucleation

Primary nucleation is the initial formation of a crystal where there are no other crystals present or where, if there are crystals present in the system, they do not have any influence on the process. This can occur in two conditions. The first is homogeneous nucleation, which is nucleation that is not influenced in any way by solids. These solids include the walls of the crystallizer vessel and particles of any foreign substance. The second category, then, is heterogeneous nucleation. This occurs when solid particles of foreign substances cause an increase in the rate of nucleation that would otherwise not be seen without the existence of these foreign particles. Homogeneous nucleation rarely occurs in practice due to the high energy necessary to begin nucleation without a solid surface to catalyse the nucleation.

Primary nucleation (both homogeneous and heterogeneous) has been modelled with the following:

$$B = \frac{dN}{dt} = k_n(c - c^*)^n$$

- B is the number of nuclei formed per unit volume per unit time.
- N is the number of nuclei per unit volume.
- k_n is a rate constant.
- c is the instantaneous solute concentration.
- c^* is the solute concentration at saturation.
- $(c–c^*)$ is also known as supersaturation.
- n is an empirical exponent that can be as large as 10, but generally ranges between 3 and 4.

Secondary Nucleation

Secondary nucleation is the formation of nuclei attributable to the influence of the existing microscopic crystals in the magma. Simply put, secondary nucleation is when crystal growth is initiated with contact of other existing crystals or "seeds." The first type of known secondary crystallization is attributable to fluid shear, the other due to collisions between already existing crystals with either a solid surface of the crystallizer or with other crystals themselves. Fluid shear nucleation occurs when liquid travels across a Crystal at a high speed, sweeping away nuclei that would otherwise be incorporated into a Crystal, causing the swept-away nuclei to become new crystals. Contact nucleation has been found to be the most effective and common method for nucleation. The benefits include the following:

- Low kinetic order and rate-proportional to supersaturation, allowing easy control without unstable operation.
- Occurs at low supersaturation, where growth rate is optimum for good quality.

- Low necessary energy at which crystals strike avoids the breaking of existing crystals into new crystals.
- The quantitative fundamentals have already been isolated and are being incorporated into practice.

The following model, although somewhat simplified, is often used to model secondary nucleation:

$$B = \frac{dN}{dt} = k_1 M_T^j (c - c^*)^b$$

- k_1 is a rate constant.
- M_T is the suspension density.
- j is an empirical exponent that can range up to 1.5, but is generally 1.
- b is an empirical exponent that can range up to 5, but is generally 2.

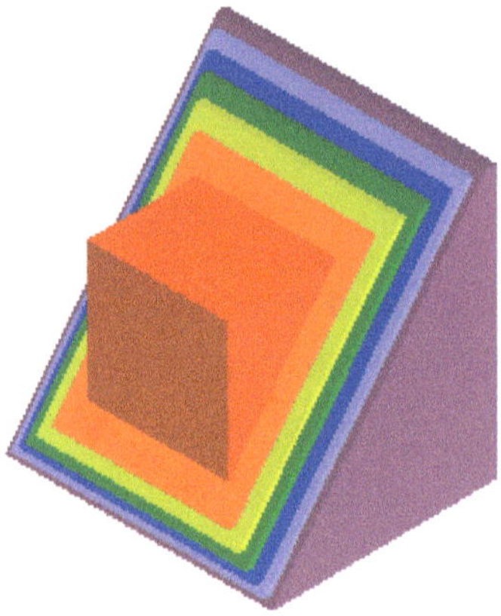

Crystal growth

Growth

Once the first small crystal, the nucleus, forms it acts as a convergence point (if unstable due to supersaturation) for molecules of solute touching – or adjacent to – the crystal so that it increases its own dimension in successive layers. The pattern of growth resembles the rings of an onion, as shown in the picture, where each colour indicates the same mass of solute; this mass creates increasingly thin layers due to the increasing surface area of the growing crystal. The supersaturated solute mass the original nucleus may *capture* in a time unit is called the *growth rate* expressed in kg/(m^2*h), and is a constant specific to the process. Growth rate is influenced by several physical factors, such as surface tension of solution, pressure, temperature, relative crystal velocity in the solution, Reynolds number, and so forth.

The main values to control are therefore:

- Supersaturation value, as an index of the quantity of solute available for the growth of the crystal;
- Total crystal surface in unit fluid mass, as an index of the capability of the solute to fix onto the crystal;
- Retention time, as an index of the probability of a molecule of solute to come into contact with an existing crystal;
- Flow pattern, again as an index of the probability of a molecule of solute to come into contact with an existing crystal (higher in laminar flow, lower in turbulent flow, but the reverse applies to the probability of contact).

The first value is a consequence of the physical characteristics of the solution, while the others define a difference between a well- and poorly designed crystallizer.

Size Distribution

The appearance and size range of a crystalline product is extremely important in crystallization. If further processing of the crystals is desired, large crystals with uniform size are important for washing, filtering, transportation, and storage. The importance lies in the fact that large crystals are easier to filter out of a solution than small crystals. Also, larger crystals have a smaller surface area to volume ratio, leading to a higher purity. This higher purity is due to less retention of mother liquor which contains impurities, and a smaller loss of yield when the crystals are washed to remove the mother liquor. The theoretical crystal size distribution can be estimated as a function of operating conditions with a fairly complicated mathematical process called population balance theory (using population balance equations).

Main Crystallization Processes

Some of the important factors influencing solubility are:

- Concentration
- Temperature
- Polarity
- Ionic Strength

So we may identify two main families of crystallization processes:

- Cooling crystallization
- Evaporative crystallization

This division is not really clear-cut, since hybrid systems exist, where cooling is performed through evaporation, thus obtaining at the same time a concentration of the solution.

A crystallization process often referred to in chemical engineering is the fractional crystallization. This is not a different process, rather a special application of one (or both) of the above.

Cooling Crystallization

Application

Most chemical compounds, dissolved in most solvents, show the so-called *direct* solubility that is, the solubility threshold increases with temperature.

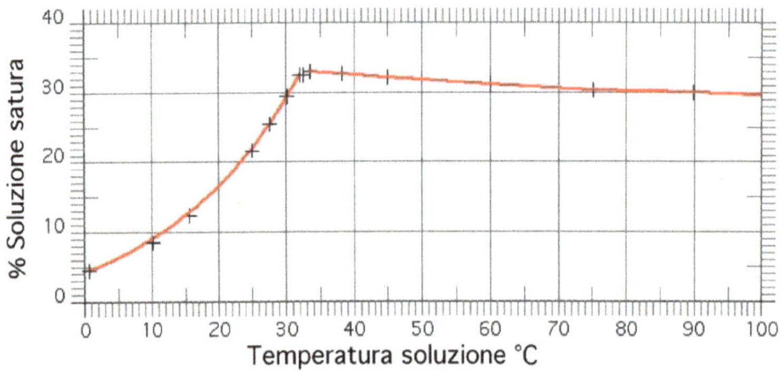

Solubility of the system $Na_2SO_4 - H_2O$

So, whenever the conditions are favourable, crystal formation results from simply cooling the solution. Here *cooling* is a relative term: austenite crystals in a steel form well above 1000 °C. An example of this crystallization process is the production of Glauber›s salt, a crystalline form of sodium sulfate. In the diagram, where equilibrium temperature is on the x-axis and equilibrium concentration (as mass percent of solute in saturated solution) in y-axis, it is clear that sulfate solubility quickly decreases below 32.5 °C. Assuming a saturated solution at 30 °C, by cooling it to 0 °C (note that this is possible thanks to the freezing-point depression), the precipitation of a mass of sulfate occurs corresponding to the change in solubility from 29% (equilibrium value at 30 °C) to approximately 4.5% (at 0 °C) – actually a larger crystal mass is precipitated, since sulfate entrains hydration water, and this has the side effect of increasing the final concentration.

There are of course limitation in the use of cooling crystallization:

- Many solutes precipitate in hydrate form at low temperatures: in the previous example this is acceptable, and even useful, but it may be detrimental when, for example, the mass of water of hydration to reach a stable hydrate crystallization form is more than the available water: a single block of hydrate solute will be formed – this occurs in the case of calcium chloride);
- Maximum supersaturation will take place in the coldest points. These may be the heat exchanger tubes which are sensitive to scaling, and heat exchange may be greatly reduced or discontinued;
- A decrease in temperature usually implies an increase of the viscosity of a solution. Too high a viscosity may give hydraulic problems, and the laminar flow thus created may affect the crystallization dynamics.
- It is of course not applicable to compounds having *reverse* solubility, a term to indicate that solubility increases with temperature decrease (an example occurs with sodium sulfate where solubility is reversed above 32.5 °C).

Cooling Crystallizers

Vertical cooling crystallizer in a beet sugar factory

The simplest cooling crystallizers are tanks provided with a mixer for internal circulation, where temperature decrease is obtained by heat exchange with an intermediate fluid circulating in a jacket. These simple machines are used in batch processes, as in processing of pharmaceuticals and are prone to scaling. Batch processes normally provide a relatively variable quality of product along the batch.

The *Swenson-Walker* crystallizer is a model, specifically conceived by Swenson Co. around 1920, having a semicylindric horizontal hollow trough in which a hollow screw conveyor or some hollow discs, in which a refrigerating fluid is circulated, plunge during rotation on a longitudinal axis. The refrigerating fluid is sometimes also circulated in a jacket around the trough. Crystals precipitate on the cold surfaces of the screw/discs, from which they are removed by scrapers and settle on the bottom of the trough. The screw, if provided, pushes the slurry towards a discharge port.

A common practice is to cool the solutions by flash evaporation: when a liquid at a given T_0 temperature is transferred in a chamber at a pressure P_1 such that the liquid saturation temperature T_1 at P_1 is lower than T_0, the liquid will release heat according to the temperature difference and a quantity of solvent, whose total latent heat of vaporization equals the difference in enthalpy. In simple words, the liquid is cooled by evaporating a part of it.

In the sugar industry vertical cooling crystallizers are used to exhaust the molasses in the last crystallization stage downstream of vacuum pans, prior to centrifugation. The massecuite enters the crystallizers at the top, and cooling water is pumped through pipes in counterflow.

Evaporative Crystallization

Another option is to obtain, at an approximately constant temperature, the precipitation of the crystals by increasing the solute concentration above the solubility threshold. To obtain this, the solute/solvent mass ratio is increased using the technique of evaporation. This process is of course insensitive to change in temperature (as long as hydration state remains unchanged).

All considerations on control of crystallization parameters are the same as for the cooling models.

Evaporative Crystallizers

Most industrial crystallizers are of the evaporative type, such as the very large sodium chloride and sucrose units, whose production accounts for more than 50% of the total world production of crystals. The most common type is the *forced circulation* (FC) model (see evaporator). A pumping device (a pump or an axial flow mixer) keeps the crystal slurry in homogeneous suspension throughout the tank, including the exchange surfaces; by controlling pump flow, control of the contact time of the crystal mass with the supersaturated solution is achieved, together with reasonable velocities at the exchange surfaces. The Oslo, mentioned above, is a refining of the evaporative forced circulation crystallizer, now equipped with a large crystals settling zone to increase the retention time (usually low in the FC) and to roughly separate heavy slurry zones from clear liquid. Evaporative crystallizers tend to yield larger average crystal size and narrows the crystal size distribution curve.

DTB Crystallizer

DTB Crystallizer

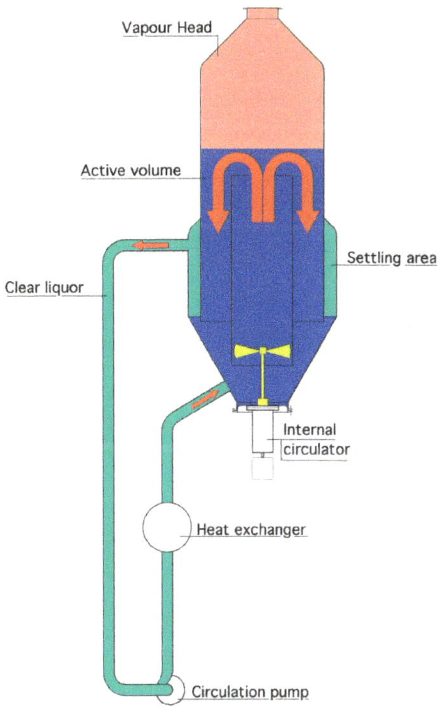

Schematic of DTB

Whichever the form of the crystallizer, to achieve an effective process control it is important to control the retention time and the crystal mass, to obtain the optimum conditions in terms of crystal specific surface and the fastest possible growth. This is achieved by a separation – to put it simply – of the crystals from the liquid mass, in order to manage the two flows in a different way. The practical way is to perform a gravity settling to be able to extract (and possibly recycle separately) the (almost) clear liquid, while managing the mass flow around the crystallizer to obtain a precise slurry density elsewhere. A typical example is the DTB (*Draft Tube and Baffle*) crystallizer, an idea of Richard Chisum Bennett (a Swenson engineer and later President of Swenson) at the end of the 1950s. The DTB crystallizer (see images) has an internal circulator, typically an axial flow mixer – yellow – pushing upwards in a draft tube while outside the crystallizer there is a settling area in an annulus; in it the exhaust solution moves upwards at a very low velocity, so that large crystals settle – and return to the main circulation – while only the fines, below a given grain size are extracted and eventually destroyed by increasing or decreasing temperature, thus creating additional supersaturation. A quasi-perfect control of all parameters is achieved as DTF crystallizers offer superior control over crystal size and characteristics.This crystallizer, and the derivative models (Krystal, CSC, etc.) could be the ultimate solution if not for a major limitation in the evaporative capacity, due to the limited diameter of the vapour head and the relatively low external circulation not allowing large amounts of energy to be supplied to the system.

DECANTATION

Decantation is a process for the separation of mixtures, by removing a layer of liquid, generally one from which a precipitate has settled. The purpose may be either to produce a clean decant, or to remove undesired liquid from the precipitate (or other layers). If the aim is to produce a clean solution, a small amount of solution must generally be left in the container, and care must be taken to prevent any precipitate from flowing with the solution out of the container.

Preliminary Processes

A mixture of an insoluble solid in liquid is allowed to stand. The solid is insoluble and settles at the bottom if kept undisturbed for some time. This process is called sedimentation. A centrifuge may be useful in decanting a solution. The centrifuge causes the precipitate to be forced to the bottom of the container; if the force is high enough, the precipitate may form a compact solid. Then the liquid can be more easily poured away, as the precipitate will tend to remain in its compressed form. Similarly a mixture of two immiscibleliquids can also be separated by decantation. For example, the oil and water extracted from fish may be decanted to obtain the oil. A mixture of kerosene and water can also be separated through decantation.

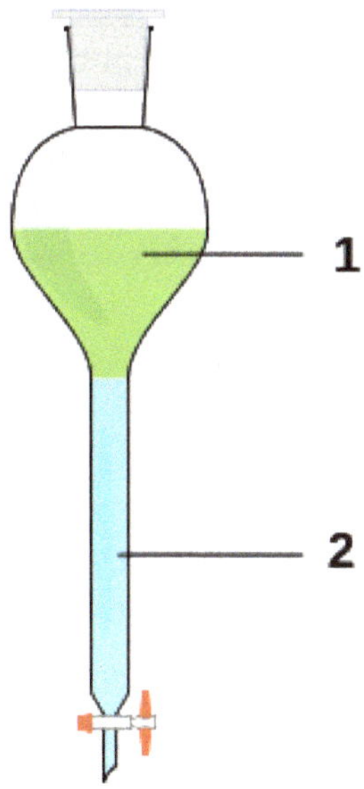

1. Upper phase
2. Lower phase

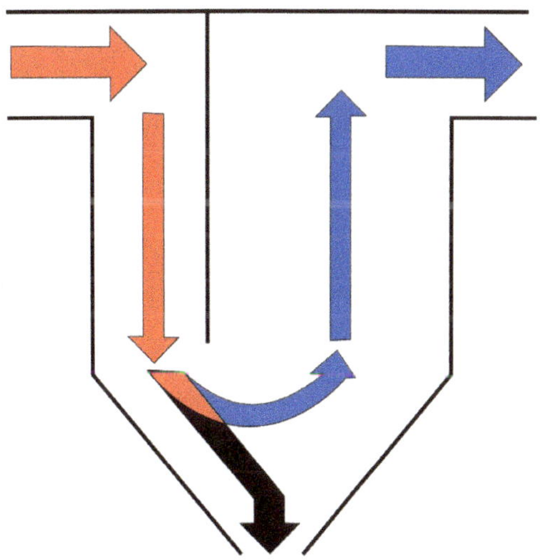

An example of decanter.

Examples

It is frequently used to purify a liquid by separating it from a suspension of insoluble particles (e.g. in red wine, where the wine is decanted from the potassium bitartrate crystals). To obtain a sample of clear water from muddy water, muddy water is left in a container until the mud settles, and then the clear water is poured into another container.

Another example of decantation is the regeneration of used chiral stationary phase (CSP). The CSP to be decanted is gently mixed in a container with a compatible solvent to form a suspension. The suspension is allowed to rest for a period of time, after which the supernatant is carefully poured off. The supernatant contains the undesirable constituents of the former suspension, while the leftover sediment in the container is clean, reusable CSP. Decantation should be done directly after sedimentation. This process is also used in erected pure iron from iron ore.

DEMISTER (VAPOR)

A demister is a device often fitted to vapor-liquid separator vessels to enhance the removal of liquid droplets entrained in a vapor stream. Demisters may be a mesh type coalescer, vane pack or other structure intended to aggregate the mist into droplets that are heavy enough to separate from the vapor stream.

Demisters can reduce the residence time required to separate a given liquid droplet size by reducing the volume and associated cost of separator equipment. Demisters are often used where vapor quality is important in regard to entrained liquids particularly where separator equipment costs are high (e.g., high pressure systems) or where space or weight savings are advantageous.

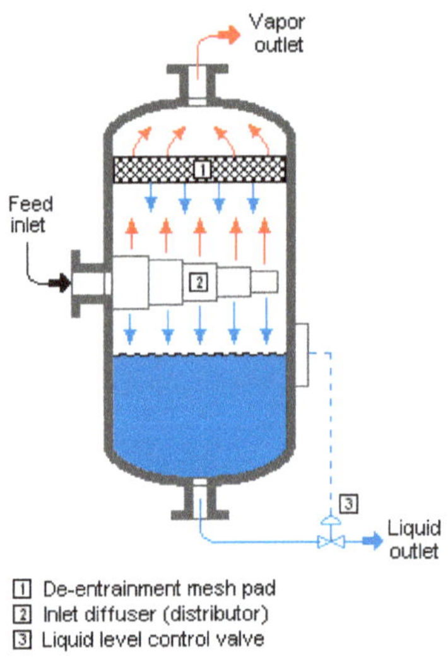

A typical, large-scale, vapor-liquid separator as used in oil refineries, petrochemical and chemical plants

For example, in the process of brine desalination on marine vessels, brine is flash heated into vapor. In flashing, vapor carries over droplets of brine which have to be separated before condensing, otherwise the distillate vapor would be contaminated with salt. This is the role of the demister. Demisted vapor condenses on tubes in the desalination plant, and product water is collected in the distillate tray.

DISTILLATION

Distillation is a process of separating the component or substances from a liquid mixture by selective evaporation and condensation. Distillation may result in essentially complete separation (nearly pure components), or it may be a partial separation that increases the concentration of selected components of the mixture. In either case the process exploits differences in the volatility of the mixture's components. In industrial chemistry, distillation is a unit operation of practically universal importance, but it is a physical separation process and not a chemical reaction.

Commercially, distillation has many applications. For example:

- In the fossil fuel industry distillation is a major class of operation in obtaining materials from crude oil for fuels and for chemical feed stocks.
- Distillation permits separation of air into its components — notably oxygen, nitrogen, and argon — for industrial use.
- In the field of industrial chemistry, large ranges of crude liquid products of chemical synthesis are distilled to separate them, either from other products, or from impurities, or from unreacted starting materials.
- Distillation of fermented products produces distilled beverages with a high alcohol content, or separates out other fermentation products of commercial value.

An installation for distillation, especially of alcohol, is a **distillery**. The distillation equipment is a **still**.

Distillation is a very old method of artificial desalination.

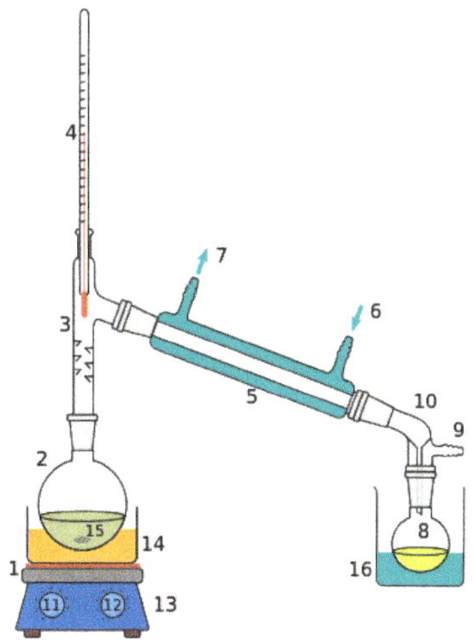

Laboratory display of distillation: **1:** A source of heat **2:** Still pot **3:** Still head **4:** Thermometer/ Boiling point temperature **5:** Condenser **6:** Cooling water in **7:** Cooling water out **8:** Distillate/ receiving flask **9:** Vacuum/gas inlet **10:** Still receiver **11:** Heat control **12:** Stirrer speed control **13:** Stirrer/heat plate **14:** Heating (Oil/sand) bath **15:** Stirring means e.g. (shown), boiling chips or mechanical stirrer **16:** Cooling bath.

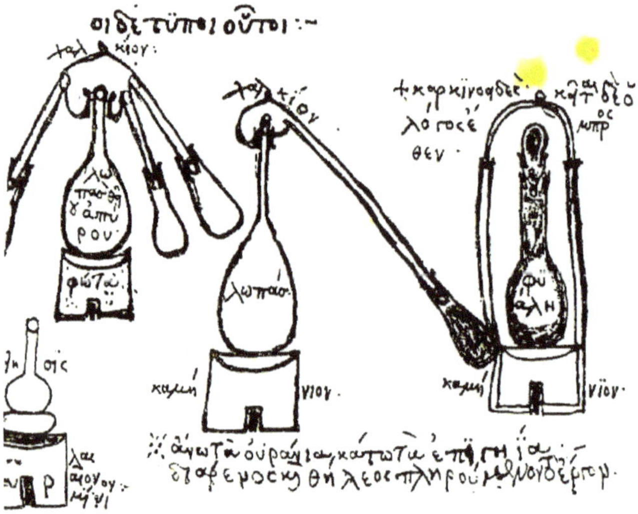

Distillation equipment used by the 3rd century Greek alchemist Zosimos of Panopolis, from the Byzantine Greek manuscript *Parisinus graces.*

Aristotle wrote about the process in his *Meteorologica* and even that "ordinary wine possesses a kind of exhalation, and that is why it gives out a flame". Later evidence of distillation comes from Greek alchemists working in Alexandria in the 1st century AD. Distilled water has been known since at least c. 200, when Alexander of Aphrodisias described the process. Work on distilling other liquids continued in early Byzantine Egypt under the Greek-Egyptian Zosimus of Panopolis. Distillation in China could have begun during the Eastern Han Dynasty (1st–2nd centuries), but archaeological evidence indicates that actual distillation of beverages began in the Jin (12th–13th centuries) and Southern Song (10th–13th centuries) dynasties. A still was found in an archaeological site in Qinglong, Hebei province dating to the 12th century. Distilled beverages were more common during the Yuan dynasty (13th–14th centuries).Arabs learned the process from the Alexandrians and used it extensively in their chemical experiments.

Clear evidence of the distillation of alcohol comes from the School of Salerno in the 12th century.Fractional distillation was developed by Tadeo Alderotti in the 13th century.

In 1500, German alchemist Hieronymus Braunschweig published *Liber de arte destillandi* (The Book of the Art of Distillation) the first book solely dedicated to the subject of distillation, followed in 1512 by a much expanded version. In 1651, John French published The Art of Distillation the first major English compendium of practice, though it has been claimed that much of it derives from Braunschweig›s work. This includes diagrams with people in them showing the industrial rather than bench scale of the operation.

Hieronymus Brunschwig's *Liber de arte Distillandi de Compositis* (Strassburg, 1512) Chemical Heritage Foundation

A retort

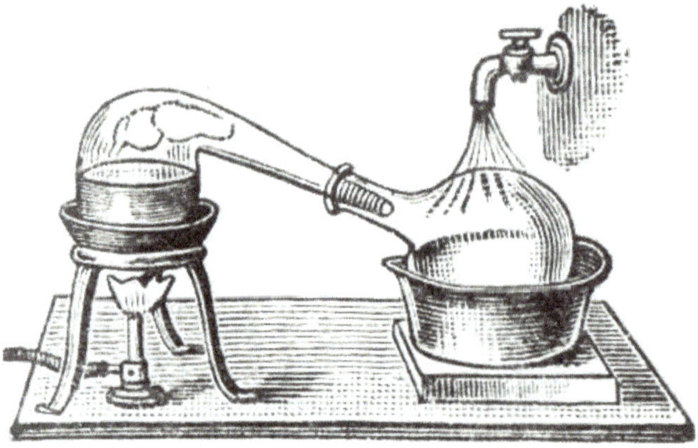

Distillation

Old Ukrainian vodka still

Simple liqueur distillation in East Timor

As alchemy evolved into the science of chemistry, vessels called retorts became used for distillations. Both alembics and retorts are forms of glassware with long necks pointing to the side at a downward angle which acted as air-cooled condensers to condense the distillate and let it drip downward for collection. Later, copper alembics were invented. Riveted joints were often kept tight by using various mixtures, for instance a dough made of rye flour. These alembics often featured a cooling system around the beak, using cold water for instance, which made the condensation of alcohol more efficient. These were called pot stills. Today, the retorts and pot stills have been largely supplanted by more efficient distillation methods in most industrial processes. However, the pot still is still widely used for the elaboration of some fine alcohols such as cognac, Scotch whisky, tequila and some vodkas. Pot stills made of various materials (wood, clay, stainless steel) are also used by bootleggers in various countries. Small pot stills are also sold for the domestic production of flower water or essential oils.

Early forms of distillation were batch processes using one vaporization and one condensation. Purity was improved by further distillation of the condensate. Greater volumes were processed by simply repeating the distillation. Chemists were reported to carry out as many as 500 to 600 distillations in order to obtain a pure compound.

In the early 19th century the basics of modern techniques including pre-heating and reflux were developed. In 1822, Anthony Perrier developed one of the first continuous stills. In 1826, Robert Stein improved that design to make his patent still. In 1830, Aeneas Coffey got a patent for improving that design. Coffey's continuous still may be regarded as the archetype of modern petrochemical units. The French engineer Armand Savalle developed his steam regulator around 1846. In 1877, Ernest Solvay was granted a U.S. Patent for a tray column for ammonia distillation and the same and subsequent years saw developments of this theme for oil and spirits.

With the emergence of chemical engineering as a discipline at the end of the 19th century, scientific rather than empirical methods could be applied. The developing petroleum industry in the early 20th century provided the impetus for the development of accurate design methods such as the McCabe–Thiele method and the Fenske equation. The availability of powerful computers has also allowed direct computer simulation of distillation columns.

Applications of Distillation

The application of distillation can roughly be divided in four groups: laboratory scale, industrial distillation, distillation of herbs for perfumery and medicinals (herbal distillate), and food processing. The latter two are distinctively different from the former two in that in the processing of beverages and herbs, the distillation is not used as a true purification method but more to transfer all volatiles from the source materials to the distillate.

The main difference between laboratory scale distillation and industrial distillation is that laboratory scale distillation is often performed batch-wise, whereas industrial distillation often occurs continuously. In batch distillation, the composition of the source material, the vapors of the distilling compounds and the distillate change during the distillation. In batch distillation, a still is charged (supplied) with a batch of feed mixture, which is then separated into its component fractions which are collected sequentially from most volatile to less volatile, with the bottoms (remaining least or non-volatile fraction) removed at the end. The still can then be recharged and the process repeated.

In continuous distillation, the source materials, vapors, and distillate are kept at a constant composition by carefully replenishing the source material and removing fractions from both vapor and liquid in the system. This results in a better control of the separation process.

Idealized Distillation Model

The boiling point of a liquid is the temperature at which the vapor pressure of the liquid equals the pressure around the liquid, enabling bubbles to form without being crushed. A special case is the normal boiling point, where the vapor pressure of the liquid equals the ambient atmospheric pressure.

It is a common misconception that in a liquid mixture at a given pressure, each component boils at the boiling point corresponding to the given pressure and the vapors of each component will collect separately and purely. This, however, does not occur even in an idealized system. Idealized models of distillation are essentially governed by Raoult's law and Dalton's law, and assume that vapor–liquid equilibria are attained.

Raoult's law states that the vapor pressure of a solution is dependent on 1) the vapor pressure of each chemical component in the solution and 2) the fraction of solution each component makes up a.k.a. the mole fraction. This law applies to ideal solutions, or solutions that have different components but whose molecular interactions are the same as or very similar to pure solutions.

Dalton's law states that the total pressure is the sum of the partial pressures of each individual component in the mixture. When a multi-component liquid is heated, the vapor pressure of each component will rise, thus causing the total vapor pressure to rise. When the total vapor pressure reaches the pressure surrounding the liquid, boiling occurs and liquid turns to gas throughout the bulk of the liquid. Note that a mixture with a given composition has one boiling point at a given pressure, when the components are mutually soluble. A mixture of constant composition does not have multiple boiling points.

An implication of one boiling point is that lighter components never cleanly "boil first". At boiling point, all volatile components boil, but for a component, its percentage in the vapor is the same as its percentage of the total vapor pressure. Lighter components have a higher partial pressure and thus are concentrated in the vapor, but heavier volatile components also have a (smaller) partial pressure and necessarily evaporate also, albeit being less concentrated in the vapor. Indeed, batch distillation and fractionation succeed by varying the composition of the mixture. In batch distillation, the batch evaporates, which changes its composition; in fractionation, liquid higher in the fractionation column contains more lights and boils at lower temperatures. Therefore, starting from a given mixture, it appears to have a **boiling range** instead of a boiling *point*, although this is because its composition changes: each intermediate mixture has its own, singular boiling point.

The idealized model is accurate in the case of chemically similar liquids, such as benzene and toluene. In other cases, severe deviations from Raoult's law and Dalton's law are observed, most famously in the mixture of ethanol and water. These compounds, when heated together, form an azeotrope, which is a composition with a boiling point higher or lower than the boiling point of each separate liquid. Virtually all liquids, when mixed and heated, will display azeotropic behaviour. Although there are computational methods that can be used to estimate the behavior of a mixture of arbitrary components, the only way to obtain accurate vapor–liquid equilibrium data is by measurement.

It is not possible to *completely* purify a mixture of components by distillation, as this would require each component in the mixture to have a zero partial pressure. If ultra-pure products are the goal, then further chemical separation must be applied. When a binary mixture is evaporated and the other component, e.g. a salt, has zero partial pressure for practical purposes, the process is simpler and is called evaporation in engineering.

Batch Distillation

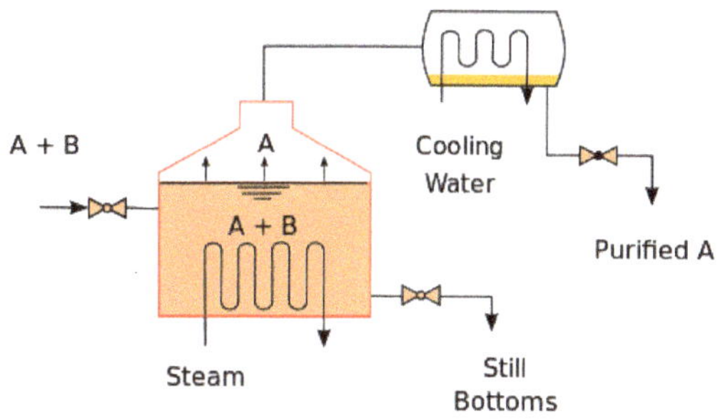

A batch still showing the separation of A and B.

Heating an ideal mixture of two volatile substances A and B (with A having the higher volatility, or lower boiling point) in a batch distillation setup (such as in an apparatus depicted in the opening figure) until the mixture is boiling results in a vapor above the liquid which contains a mixture of A and B. The ratio between A and B in the vapor will be different from the ratio in the liquid: the ratio in the liquid will be determined by how the original mixture was prepared, while the ratio in the vapor will be enriched in the more volatile compound, A (due to Raoult's Law, see above). The vapor goes through the condenser and is removed from the system. This in turn means that the ratio of compounds in the remaining liquid is now different from the initial ratio (i.e., more enriched in B than the starting liquid).

The result is that the ratio in the liquid mixture is changing, becoming richer in component B. This causes the boiling point of the mixture to rise, which in turn results in a rise in the temperature in the vapor, which results in a changing ratio of A : B in the gas phase (as distillation continues, there is an increasing proportion of B in the gas phase). This results in a slowly changing ratio A : B in the distillate.

If the difference in vapor pressure between the two components A and B is large (generally expressed as the difference in boiling points), the mixture in the beginning of the distillation is highly enriched in component A, and when component A has distilled off, the boiling liquid is enriched in component B.

Continuous Distillation

Continuous distillation is an ongoing distillation in which a liquid mixture is continuously (without interruption) fed into the process and separated fractions are removed continuously as output streams occur over time during the operation. Continuous distillation produces a minimum of two output fractions, including at least one volatile distillate fraction, which has boiled and been separately captured as a vapor, and then condensed to a liquid. There is always a bottoms (or residue) fraction, which is the least volatile residue that has not been separately captured as a condensed vapor.

Continuous distillation differs from batch distillation in the respect that concentrations should not change over time. Continuous distillation can be run at a steady state for an arbitrary amount of time. For any source material of specific composition, the main variables that affect the purity of products in continuous distillation are the reflux ratio and the number of theoretical equilibrium stages, in practice determined by the number of trays or the height of packing. Reflux is a flow from the condenser back to the column, which generates a recycle that allows a better separation with a given number of trays. Equilibrium stages are ideal steps where compositions achieve vapor–liquid equilibrium, repeating the separation process and allowing better separation given a reflux ratio. A column with a high reflux ratio may have fewer stages, but it refluxes a large amount of liquid, giving a wide column with a large holdup. Conversely, a column with a low reflux ratio must have a large number of stages, thus requiring a taller column.

General Improvements

Both batch and continuous distillations can be improved by making use of a fractionating column on top of the distillation flask. The column improves separation by providing a larger surface area for the vapor and condensate to come into contact. This helps it remain at equilibrium for as long as possible. The column can even consist of small subsystems ('trays' or 'dishes') which all contain an enriched, boiling liquid mixture, all with their own vapor–liquid equilibrium.

There are differences between laboratory-scale and industrial-scale fractionating columns, but the principles are the same. Examples of laboratory-scale fractionating columns (in increasing efficiency) include

- Air condenser
- Vigreux column (usually laboratory scale only)
- Packed column (packed with glass beads, metal pieces, or other chemically inert material)
- Spinning band distillation system.

LABORATORY SCALE DISTILLATION

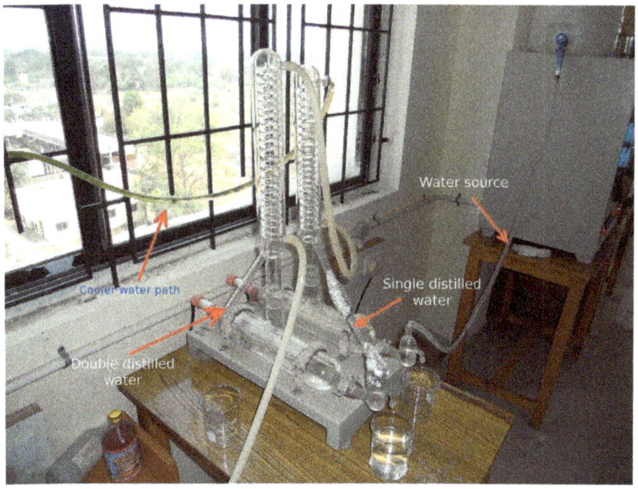

Typical laboratory distillation unit

Laboratory scale distillations are almost exclusively run as batch distillations. The device used in distillation, sometimes referred to as a *still*, consists at a minimum of a **reboiler** or *pot* in which the source material is heated, a **condenser** in which the heated vapour is cooled back to the liquid state, and a **receiver** in which the concentrated or purified liquid, called the **distillate**, is collected. Several laboratory scale techniques for distillation exist.

Simple Distillation

In **simple distillation**, the vapor is immediately channeled into a condenser. Consequently, the distillate is not pure but rather its composition is identical to the composition of the vapors at the given temperature and pressure. That concentration follows Raoult's law.

As a result, simple distillation is effective only when the liquid boiling points differ greatly (rule of thumb is 25 °C) or when separating liquids from non-volatile solids or oils. For these cases, the vapor pressures of the components are usually different enough that the distillate may be sufficiently pure for its intended purpose.

Fractional Distillation

For many cases, the boiling points of the components in the mixture will be sufficiently close that Raoult's law must be taken into consideration. Therefore, **fractional distillation** must be used in order to separate the components by repeated vaporization-condensation cycles within a packed fractionating column. This separation, by successive distillations, is also referred to as **rectification**.

As the solution to be purified is heated, its vapors rise to the fractionating column. As it rises, it cools, condensing on the condenser walls and the surfaces of the packing material. Here, the condensate continues to be heated by the rising hot vapors; it vaporizes once more. However, the composition of the fresh vapors are determined once again by Raoult's law. Each vaporization-condensation cycle (called a *theoretical plate*) will yield a purer solution of the more volatile component. In reality, each cycle at a given temperature does not occur at exactly the same position in the fractionating column; *theoretical plate* is thus a concept rather than an accurate description.

More theoretical plates lead to better separations. A spinning band distillation system uses a spinning band of Teflon or metal to force the rising vapors into close contact with the descending condensate, increasing the number of theoretical plates.

Steam Distillation

Like vacuum distillation, **steam distillation** is a method for distilling compounds which are heat-sensitive. The temperature of the steam is easier to control than the surface of a heating element, and allows a high rate of heat transfer without heating at a very high temperature. This process involves bubbling steam through a heated mixture of the raw material. By Raoult›s law, some of the target compound will vaporize (in accordance with its partial pressure). The vapor mixture is cooled and condensed, usually yielding a layer of oil and a layer of water.

Steam distillation of various aromatic herbs and flowers can result in two products; an essential oil as well as a watery herbal distillate. The essential oils are often used in perfumery and aromatherapy while the watery distillates have many applications in aromatherapy, food processing and skin care.

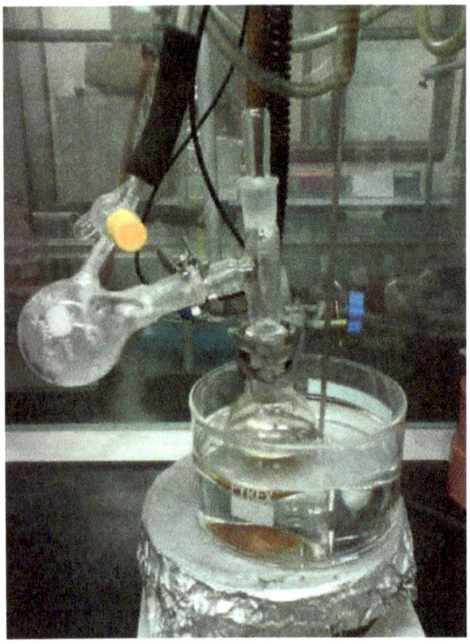

Dimethyl sulfoxide usually boils at 189 °C. Under a vacuum, it distills off into the receiver at only 70 °C.

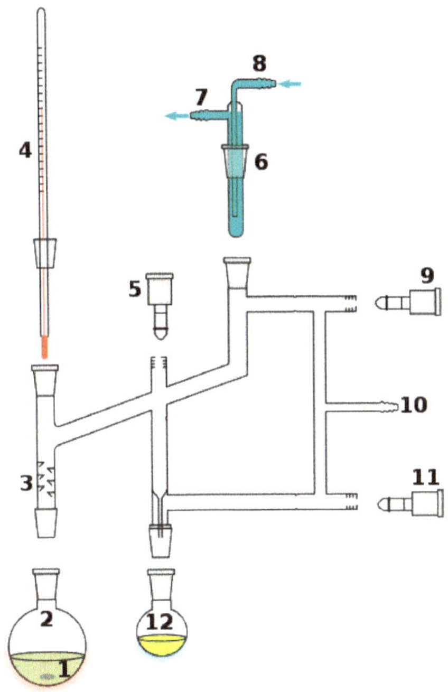

Perkin triangle distillation setup

1: Stirrer bar/anti-bumping granules **2:** Still pot **3:** Fractionating column **4:** Thermometer/ Boiling point temperature **5:** Teflon tap 1 **6:** Cold finger **7:** Cooling water out **8:** Cooling water in **9:** Teflon tap 2 **10:** Vacuum/gas inlet **11:** Teflon tap 3 **12:** Still receiver

Vacuum Distillation

Some compounds have very high boiling points. To boil such compounds, it is often better to lower the pressure at which such compounds are boiled instead of increasing the temperature. Once the pressure is lowered to the vapor pressure of the compound (at the given temperature), boiling and the rest of the distillation process can commence. This technique is referred to as **vacuum distillation** and it is commonly found in the laboratory in the form of the rotary evaporator.

This technique is also very useful for compounds which boil beyond their decomposition temperature at atmospheric pressure and which would therefore be decomposed by any attempt to boil them under atmospheric pressure.

Molecular distillation is vacuum distillation below the pressure of 0.01 torr. 0.01 torr is one order of magnitude above high vacuum, where fluids are in the free molecular flow regime, i.e. the mean free path of molecules is comparable to the size of the equipment. The gaseous phase no longer exerts significant pressure on the substance to be evaporated, and consequently, rate of evaporation no longer depends on pressure. That is, because the continuum assumptions of fluid dynamics no longer apply, mass transport is governed by molecular dynamics rather than fluid dynamics. Thus, a short path between the hot surface and the cold surface is necessary, typically by suspending a hot plate covered with a film of feed next to a cold plate with a line of sight in between. Molecular distillation is used industrially for purification of oils.

Air-Sensitive Vacuum Distillation

Some compounds have high boiling points as well as being air sensitive. A simple vacuum distillation system as exemplified above can be used, whereby the vacuum is replaced with an inert gas after the distillation is complete. However, this is a less satisfactory system if one desires to collect fractions under a reduced pressure. To do this a "cow" or "pig" adaptor can be added to the end of the condenser, or for better results or for very air sensitive compounds a Perkin triangle apparatus can be used.

The Perkin triangle, has means via a series of glass or Teflon taps to allows fractions to be isolated from the rest of the still, without the main body of the distillation being removed from either the vacuum or heat source, and thus can remain in a state of reflux. To do this, the sample is first isolated from the vacuum by means of the taps, the vacuum over the sample is then replaced with an inert gas (such as nitrogen or argon) and can then be stoppered and removed. A fresh collection vessel can then be added to the system, evacuated and linked back into the distillation system via the taps to collect a second fraction, and so on, until all fractions have been collected.

Short Path Distillation

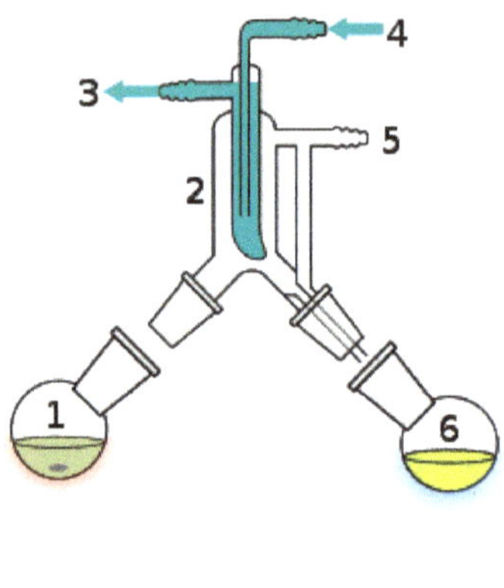

Short path vacuum distillation apparatus with vertical condenser (cold finger), to minimize the distillation path; **1:** Still pot with stirrer bar/anti-bumping granules **2:** Cold finger – bent to direct condensate **3:** Cooling water out **4:** cooling water in **5:** Vacuum/gas inlet **6:** Distillate flask/distillate.

Short path distillation is a distillation technique that involves the distillate travelling a short distance, often only a few centimeters, and is normally done at reduced pressure. A classic example would be a distillation involving the distillate travelling from one glass bulb to another, without the need for a condenser separating the two chambers. This technique is often used for compounds which are unstable at high temperatures or to purify small amounts of compound. The advantage is that the heating temperature can be considerably lower (at reduced pressure) than the boiling point of the liquid at standard pressure, and the distillate only has to travel a short distance before condensing. A short path ensures that little compound is lost on the sides of the apparatus. The Kugelrohr is a kind of a short path distillation apparatus which often contain multiple chambers to collect distillate fractions.

Zone Distillation

Zone distillation is a distillation process in long container with partial melting of refined matter in moving liquid zone and condensation of vapor in the solid phase at condensate pulling in cold area. The process is worked in theory. When zone heater is moving from the top to the bottom of the container then solid condensate with irregular impurity distribution is forming. Then most pure part of the condensate may be extracted as product. The process may be iterated many times by moving (without turnover) the received condensate to the bottom part of the container on the place of refined matter. The irregular impurity distribution in the condensate (that is efficiency of purification) increases with number of repetitions of the process. Zone distillation is a distillation analog of zone recrystallization. Impurity distribution in the condensate is described by known equations of zone recrystallization with various numbers of iteration of process – with replacement distribution efficient k of crystallization on separation factor α of distillation.

Other Types

- The process of reactive distillation involves using the reaction vessel as the still. In this process, the product is usually significantly lower-boiling than its reactants. As the product is formed from the reactants, it is vaporized and removed from the reaction mixture. This technique is an example of a continuous vs. a batch process; advantages include less downtime to charge the reaction vessel with starting material, and less workup. Distillation "over a reactant" could be classified as a reactive distillation. It is typically used to remove volatile impurity from the distallation feed. For example, a little lime may be added to remove carbon dioxide from water followed by a second distillation with a little sulfuric acid added to remove traces of ammonia.
- Catalytic distillation is the process by which the reactants are catalyzed while being distilled to continuously separate the products from the reactants. This method is used to assist equilibrium reactions reach completion.
- Pervaporation is a method for the separation of mixtures of liquids by partial vaporization through a non-porous membrane.
- Extractive distillation is defined as distillation in the presence of a miscible, high boiling, relatively non-volatile component, the solvent, that forms no azeotrope with the other components in the mixture.

- Flash evaporation (or partial evaporation) is the partial vaporization that occurs when a saturated liquid stream undergoes a reduction in pressure by passing through a throttling valve or other throttling device. This process is one of the simplest unit operations, being equivalent to a distillation with only one equilibrium stage.
- Codistillation is distillation which is performed on mixtures in which the two compounds are not miscible. In the laboratory, the Dean-Stark apparatus is used for this purpose to remove water from synthesis products.

The unit process of evaporation may also be called "distillation":

- In rotary evaporation a vacuum distillation apparatus is used to remove bulk solvents from a sample. Typically the vacuum is generated by a water aspirator or a membrane pump.
- In a kugelrohr a short path distillation apparatus is typically used (generally in combination with a (high) vacuum) to distill high boiling (> 300 °C) compounds. The apparatus consists of an oven in which the compound to be distilled is placed, a receiving portion which is outside of the oven, and a means of rotating the sample. The vacuum is normally generated by using a high vacuum pump.

Other uses:

- Dry distillation or destructive distillation, despite the name, is not truly distillation, but rather a chemical reaction known as pyrolysis in which solid substances are heated in an inert or reducing atmosphere and any volatile fractions, containing high-boiling liquids and products of pyrolysis, are collected. The destructive distillation of wood to give methanol is the root of its common name – *wood alcohol.*
- Freeze distillation is an analogous method of purification using freezing instead of evaporation. It is not truly distillation, but a recrystallization where the product is the mother liquor, and does not produce products equivalent to distillation. This process is used in the production of ice beer and ice wine to increase ethanol and sugar content, respectively. It is also used to produce applejack. Unlike distillation, freeze distillation concentrates poisonous congeners rather than removing them; As a result, many countries prohibit such applejack as a health measure. However, reducing methanol with the absorption of 4A molecular sieve is a practical method for production. Also, distillation by evaporation can separate these since they have different boiling points.

Azeotropic Distillation

Interactions between the components of the solution create properties unique to the solution, as most processes entail nonideal mixtures, where Raoult's law does not hold. Such interactions can result in a constant-boiling **azeotrope** which behaves as if it were a pure compound (i.e., boils at a single temperature instead of a range). At an azeotrope, the solution contains the given component in the same proportion as the vapor, so that evaporation does not change the purity, and distillation does not effect separation. For example, ethyl alcohol and water form an azeotrope of 95.6% at 78.1 °C.

If the azeotrope is not considered sufficiently pure for use, there exist some techniques to break the azeotrope to give a pure distillate. This set of techniques are known as **azeotropic distillation**. Some techniques achieve this by "jumping" over the azeotropic composition (by adding another component to create a new azeotrope, or by varying the pressure). Others work by chemically or physically removing or sequestering the impurity. For example, to purify ethanol beyond 95%, a drying agent (or desiccant, such as potassium carbonate) can be added to convert the soluble water into insoluble water of crystallization. Molecular sieves are often used for this purpose as well.

Immiscible liquids, such as water and toluene, easily form azeotropes. Commonly, these azeotropes are referred to as a low boiling azeotrope because the boiling point of the azeotrope is lower than the boiling point of either pure component. The temperature and composition of the azeotrope is easily predicted from the vapor pressure of the pure components, without use of Raoult's law. The azeotrope is easily broken in a distillation set-up by using a liquid–liquid separator (a decanter) to separate the two liquid layers that are condensed overhead. Only one of the two liquid layers is refluxed to the distillation set-up.

High boiling azeotropes, such as a 20 weight percent mixture of hydrochloric acid in water, also exist. As implied by the name, the boiling point of the azeotrope is greater than the boiling point of either pure component.

To break azeotropic distillations and cross distillation boundaries, such as in the DeRosier Problem, it is necessary to increase the composition of the light key in the distillate.

Breaking an Azeotrope with Unidirectional Pressure Manipulation

The boiling points of components in an azeotrope overlap to form a band. By exposing an azeotrope to a vacuum or positive pressure, it's possible to bias the boiling point of one component away from the other by exploiting the differing vapour pressure curves of each; the curves may overlap at the azeotropic point, but are unlikely to be remain identical further along the pressure axis either side of the azeotropic point. When the bias is great enough, the two boiling points no longer overlap and so the azeotropic band disappears.

This method can remove the need to add other chemicals to a distillation, but it has two potential drawbacks.

Under negative pressure, power for a vacuum source is needed and the reduced boiling points of the distillates requires that the condenser be run cooler to prevent distillate vapours being lost to the vacuum source. Increased cooling demands will often require additional energy and possibly new equipment or a change of coolant.

Alternatively, if positive pressures are required, standard glassware can not be used, energy must be used for pressurization and there is a higher chance of side reactions occurring in the distillation, such as decomposition, due to the higher temperatures required to effect boiling.

A unidirectional distillation will rely on a pressure change in one direction, either positive or negative.

Pressure-Swing Distillation

Pressure-swing distillation is essentially the same as the unidirectional distillation used to break azeotropic mixtures, but here both positive and negative pressures may be employed.

This improves the selectivity of the distillation and allows a chemist to optimize distillation by avoiding extremes of pressure and temperature that waste energy. This is particularly important in commercial applications.

One example of the application of pressure-swing distillation is during the industrial purification of ethyl acetate after its catalytic synthesis from ethanol.

Industrial Distillation

Typical industrial distillation towers

Large scale **industrial distillation** applications include both batch and continuous fractional, vacuum, azeotropic, extractive, and steam distillation. The most widely used industrial applications of continuous, steady-state fractional distillation are in petroleum refineries, petrochemical and chemical plants and natural gas processing plants.

To control and optimize such industrial distillation, a standardized laboratory method, ASTM D86, is established. This test method extends to the atmospheric distillation of petroleum products using a laboratory batch distillation unit to quantitatively determine the boiling range characteristics of petroleum products.

Industrial distillation is typically performed in large, vertical cylindrical columns known as **distillation towers** or **distillation columns** with diameters ranging from about 65 centimeters to 16 meters and heights ranging from about 6 meters to 90 meters or more. When the process feed has a diverse composition, as in distilling crude oil, liquid outlets at intervals up the column allow for the withdrawal of different *fractions* or products having different boiling points or boiling ranges. The "lightest" products (those with the lowest boiling point) exit from the top of the columns and the "heaviest" products (those with the highest boiling point) exit from the bottom of the column and are often called the **bottoms**.

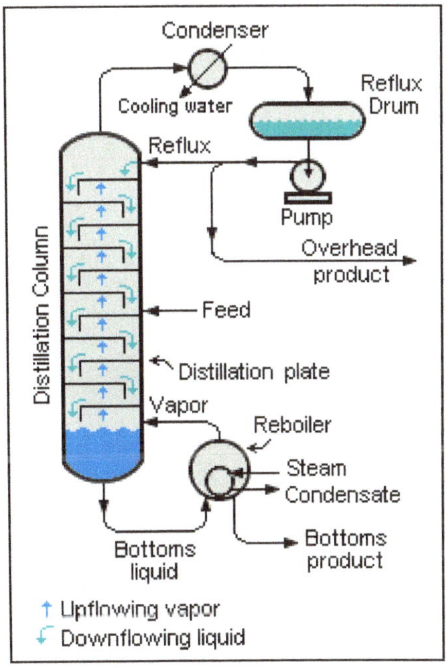

Diagram of a typical industrial distillation tower

Industrial towers use reflux to achieve a more complete separation of products. Reflux refers to the portion of the condensed overhead liquid product from a distillation or fractionation tower that is returned to the upper part of the tower as shown in the schematic diagram of a typical, large-scale industrial distillation tower. Inside the tower, the downflowing reflux liquid provides cooling and condensation of the upflowing vapors thereby increasing the efficiency of the distillation tower. The more reflux that is provided for a given number of theoretical plates, the better the tower's separation of lower boiling materials from higher boiling materials. Alternatively, the more reflux that is provided for a given desired separation, the fewer the number of theoretical plates required. Chemical engineers must choose what combination of reflux rate and number of plates is both economically and physically feasible for the products purified in the distillation column.

Such industrial fractionating towers are also used in cryogenic air separation, producing liquid oxygen, liquid nitrogen, and high purity argon. Distillation of chlorosilanes also enables the production of high-purity silicon for use as a semiconductor.

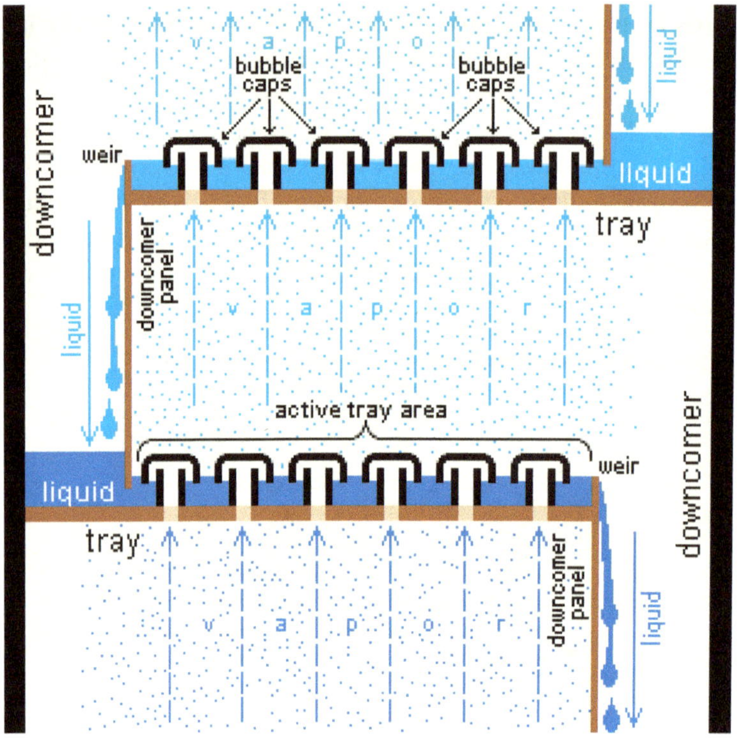

Section of an industrial distillation tower showing detail of trays with bubble caps

Design and operation of a distillation tower depends on the feed and desired products. Given a simple, binary component feed, analytical methods such as the McCabe–Thiele method or the Fenske equation can be used. For a multi-component feed, simulation models are used both for design and operation. Moreover, the efficiencies of the vapor–liquid contact devices (referred to as “plates” or “trays”) used in distillation towers are typically lower than that of a theoretical 100% efficient equilibrium stage. Hence, a distillation tower needs more trays than the number of theoretical vapor–liquid equilibrium stages. A variety of models have been postulated to estimate tray efficiencies.

In modern industrial uses, a packing material is used in the column instead of trays when low pressure drops across the column are required. Other factors that favor packing are: vacuum systems, smaller diameter columns, corrosive systems, systems prone to foaming, systems requiring low liquid holdup, and batch distillation. Conversely, factors that favor plate columns are: presence of solids in feed, high liquid rates, large column diameters, complex columns, columns with wide feed composition variation, columns with a chemical reaction, absorption columns, columns limited by foundation weight tolerance, low liquid rate, large turn-down ratio and those processes subject to process surges.

Large-scale, industrial vacuum distillation column

This packing material can either be random dumped packing (1–3" wide) such as Raschig rings or structured sheet metal. Liquids tend to wet the surface of the packing and the vapors pass across this wetted surface, where mass transfer takes place. Unlike conventional tray distillation in which every tray represents a separate point of vapor–liquid equilibrium, the vapor–liquid equilibrium curve in a packed column is continuous. However, when modeling packed columns, it is useful to compute a number of "theoretical stages" to denote the separation efficiency of the packed column with respect to more traditional trays. Differently shaped packings have different surface areas and void space between packings. Both of these factors affect packing performance.

Another factor in addition to the packing shape and surface area that affects the performance of random or structured packing is the liquid and vapor distribution entering the packed bed. The number of theoretical stages required to make a given separation is calculated using a specific vapor to liquid ratio. If the liquid and vapor are not evenly distributed across the superficial tower area as it enters the packed bed, the liquid to vapor ratio will not be correct in the packed bed and the required separation will not be achieved. The packing will appear to not be working properly. The height equivalent to a theoretical plate (HETP) will be greater than expected. The problem is not the packing itself but the mal-distribution of the fluids entering the packed bed. Liquid mal-distribution is more frequently the problem than vapor. The design of the liquid distributors used to introduce the feed and reflux to a packed bed is critical to making the packing perform to it maximum efficiency. Methods of evaluating the effectiveness of a liquid distributor to evenly distribute the liquid entering a packed bed can be found in references. Considerable work as been done on this topic by Fractionation Research, Inc. (commonly known as FRI).

Multi-Effect Distillation

The goal of multi-effect distillation is to increase the energy efficiency of the process, for use in desalination, or in some cases one stage in the production of ultrapure water. The number of effects is inversely proportional to the kW·h/m^3 of water recovered figure, and refers to the volume of water recovered per unit of energy compared with single-effect distillation. One effect is roughly 636 kW·h/m^3.

- Multi-stage flash distillation Can achieve more than 20 effects with thermal energy input, as mentioned in the article.
- Vapor compression evaporation Commercial large-scale units can achieve around 72 effects with electrical energy input, according to manufacturers.

There are many other types of multi-effect distillation processes, including one referred to as simply multi-effect distillation (MED), in which multiple chambers, with intervening heat exchangers, are employed.

Distillation in Food Processing

Distilled Beverages

Carbohydrate-containing plant materials are allowed to ferment, producing a dilute solution of ethanol in the process. Spirits such as whiskey and rum are prepared by distilling these dilute solutions of ethanol. Components other than ethanol, including water, esters, and other alcohols, are collected in the condensate, which account for the flavor of the beverage. Some of these beverages are then stored in barrels or other containers to acquire more flavor compounds and characteristic flavors.

Gallery

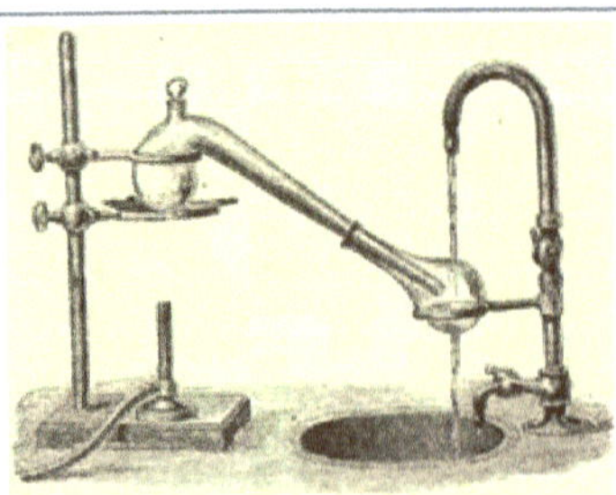	Chemistry in its beginnings used retorts as laboratory equipment exclusively for distillation processes.

A simple set-up to distill dry and oxygen-free toluene.

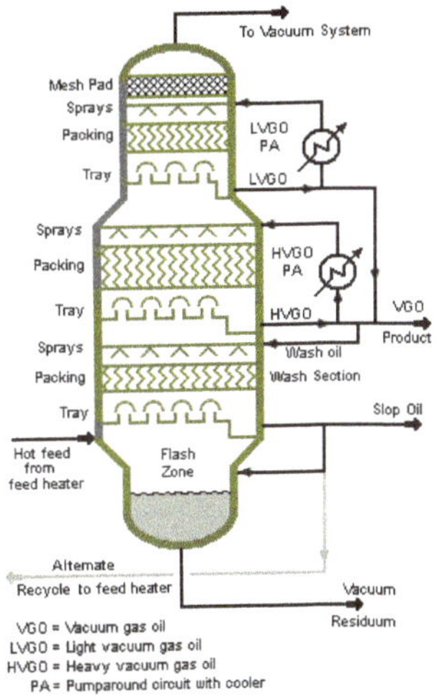

Diagram of an industrial-scale vacuum distillation column as commonly used in oil refineries

A rotary evaporator is able to distill solvents more quickly at lower temperatures through the use of a vacuum.

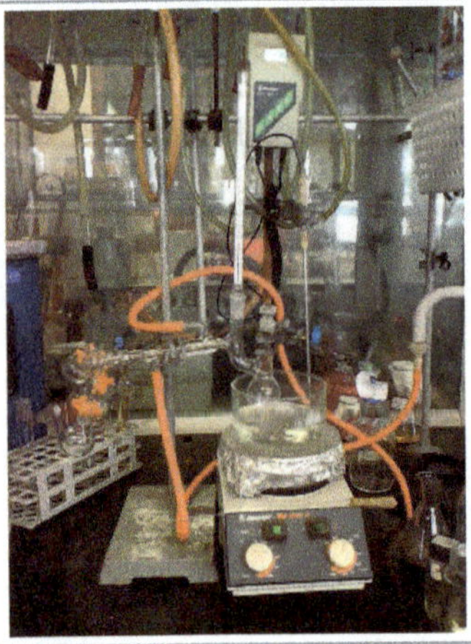	Distillation using semi-microscale apparatus. The jointless design eliminates the need to fit pieces together. The pear-shaped flask allows the last drop of residue to be removed, compared with a similarly-sized round-bottom flask The small holdup volume prevents losses. A pig is used to channel the various distillates into three receiving flasks. If necessary the distillation can be carried out under vacuum using the vacuum adapter at the pig.

REFERENCES

1. Tavare, N.S. (1995). Industrial Crystallization Plenum Press, New York
2. McCabe & Smith (2000). Unit Operations of Chemical Engineering' McGraw-Hill, New York
3. "Crystallization". www.reciprocalnet.org. Retrieved 2017-01-03.
4. "Submerge Circulating Crystallizers - Thermal Kinetics Engineering, PLLC". Thermal Kinetics Engineering, PLLC. Retrieved 2017-01-03.
5. "Draft Tube Baffle (DTB) Crystallizer - Swenson Technology". Swenson Technology. Retrieved 2017-01-03.
6. Harwood, Laurence M.; Moody, Christopher J. (1989). Experimental organic chemistry: Principles and Practice (Illustrated ed.). Oxford: Blackwell Scientific Publications. pp. 141–143. ISBN 978-0-632-02017-1.
7. E. Gildemeister; Fr. Hoffman; translated by Edward Kremers (1913). The Volatile Oils. 1. New York: Wiley. p. 203.
8. Bryan H. Bunch; Alexander Hellemans (2004). The History of Science and Technology. Houghton Mifflin Harcourt. p. 88. ISBN 0-618-22123-9.
9. Marcelin Berthelot Collection des anciens alchimistes grecs (3 vol., Paris, 1887–1888, p.161)
10. Berthelot, M. P. E. M. (1893). "The Discovery of Alcohol and Distillation". The Popular Science Monthly. XLIII: 85–94.
11. Forbes, Robert James (1970). A short history of the art of distillation: from the beginnings up to the death of Cellier Blumenthal. BRILL. pp. 57, 89. ISBN 978-90-04-00617-1. Retrieved 29 June 2010.
12. Taylor, F. (1945). "The evolution of the still". Annals of Science. 5 (3): 185. doi:10.1080/00033794500201451.
13. Haw, Stephen G. (2012). "Wine, women and poison". Marco Polo in China. Routledge. pp. 147–148. ISBN 978-1-134-27542-7. The earliest possible period

seems to be the Eastern Han dynasty ... the most likely period for the beginning of true distillation of spirits for drinking in China is during the Jin and Southern Song dynasties

14. Sarton, George (1975). Introduction to the history of science. R. E. Krieger Pub. Co. p. 145. ISBN 0-88275-172-7.
15. Holmyard, Eric John (1990). Alchemy. Courier Dover Publications. p. 53. ISBN 0-486-26298-7.
16. Magnum Opus Hermetic Sourceworks Series. Alchemywebsite.com. Retrieved on 2014-04-20.
17. Industrial Engineering Chemistry (1936) p. 677
18. Sealing Technique, accessed 16 November 2006.
19. Traditional Alembic Pot Still, accessed 16 November 2006.
20. D. F. Othmer (1982) "Distillation – Some Steps in its Development", in W. F. Furter (ed) A Century of Chemical Engineering ISBN 0-306-40895-3
21. GB 5974, Coffey, A., "Apparatus for Brewing and Distilling", published 5 August 1830, issued 5 February 1831; image
22. Forbes, Robert J. A Short History of the Art of Distillation.
23. US 198699, Solvay, Ernest, "Improvement in the Ammonia-Soda Manufacture", published 2 June 1876, issued 25 December 1877
24. ST07 Separation of liquid–liquid mixtures (solutions), DIDAC by IUPAC
25. Perry, Robert H.; Green, Don W. (1984). Perry's Chemical Engineers' Handbook (6th ed.). McGraw-Hill. ISBN 0-07-049479-7.
26. Fractional Distillation. fandm.edu
27. Spinning Band Distillation. B/R Instrument Corporation (accessed 8 September 2006)
28. Harwood & Moody 1989, pp. 151–153
29. Vogel's 5th ed.
30. Harwood & Moody 1989, p. 150
31. Kravchenko, A. I. (2011). "Zone distillation: a new method of refining". Problems of atomic science and technology (in Russian). 6 (19): 24–26.
32. Kravchenko, A. I. (2014). "Zone distillation: justification". Problems of atomic science and technology. 1 (20): 64–65.
33. Kravchenko, A. I. (2014). "Design of advanced processes of zone distillation". Perspectivnye materialy (in Russian) (7): 68–72.
34. Study on Method of Decreasing Methanol in Apple Pomace Spirit.
35. Kister, Henry Z. (1992). Distillation Design (1st ed.). McGraw-Hill. ISBN 0-07-034909-6.
36. Seader, J. D.; Henley, Ernest J. (1998). Separation Process Principles. New York: Wiley. ISBN 0-471-58626-9.

37. Energy Institute website page. Resources.schoolscience.co.uk. Retrieved on 2014-04-20.
38. Moore, F., Rukovena, F. (August 1987) Random Packing, Vapor and Liquid Distribution: Liquid and gas distribution in commercial packed towers, Chemical Plants & Processing, Edition Europe, pp. 11–15
39. Spiegel, L (2006). "A new method to assess liquid distributor quality". Chemical Engineering and Processing. 45 (11): 1011. doi:10.1016/j.cep.2006.05.003.
40. Kunesh, John G.; Lahm, Lawrence; Yanagi, Takashi (1987). "Commercial scale experiments that provide insight on packed tower distributors". Industrial & Engineering Chemistry Research. 26 (9): 1845. doi:10.1021/ie00069a021.

CHAPTER 5

Drying Techniques

Chapter Contents

- **Introduction**
- **Electrophoresis**
- **Mineral Processing**
- **Elutriation**
- **Evaporation**

INTRODUCTION

Drying is a mass transfer process consisting of the removal of water or another solvent by evaporation from a solid, semi-solid or liquid. This process is often used as a final production step before selling or packaging products. To be considered "dried", the final product must be solid, in the form of a continuous sheet (e.g., paper), long pieces (e.g., wood), particles (e.g., cereal grains or corn flakes) or powder (e.g., sand, salt, washing powder, milk powder). A source of heat and an agent to remove the vapor produced by the process are often involved. In bioproducts like food, grains, and pharmaceuticals like vaccines, the solvent to be removed is almost invariably water. Desiccation may be synonymous with drying or considered an extreme form of drying.

In the most common case, a gas stream, e.g., air, applies the heat by convection and carries away the vapor as humidity. Other possibilities are vacuum drying, where heat is supplied by conduction or radiation (or microwaves), while the vapor thus produced is removed by the vacuum system. Another indirect technique is drum drying (used, for instance, for manufacturing potato flakes), where a heated surface is used to provide the energy, and aspirators draw the vapor outside the room. In contrast, the mechanical extraction of the solvent, e.g., water, by centrifugation, is not considered "drying" but rather "draining".

Drying Mechanism

In some products having a relatively high initial moisture content, an initial linear reduction of the average product moisture content as a function of time may be observed for a limited time, often known as a "constant drying rate period". Usually, in this period, it is surface moisture outside individual particles that is being removed. The drying rate during this period is mostly dependent on the rate of heat transfer to the material being dried. Therefore, the maximum achievable drying rate is considered to be heat-transfer limited. If drying is continued, the slope of the curve, the drying rate, becomes less steep (falling rate period) and eventually tends to nearly horizontal at very long times. The product moisture content is then constant at the "equilibrium moisture content", where it is, in practice, in equilibrium with the dehydrating medium. In the falling-rate period, water migration from the product interior to the surface is mostly by molecular diffusion, i,e. the water flux is proportional to the moisture content gradient. This means that water moves from zones with higher moisture content to zones with lower values, a phenomenon explained by the second law of thermodynamics. If water removal is considerable, the products usually undergo shrinkage and deformation, except in a well-designed freeze-drying process. The drying rate in the falling-rate period is controlled by the rate of removal of moisture or solvent from the interior of the solid being dried and is referred to as being "mass-transfer limited". This is widely noticed in hygroscopic products such as fruits and vegetables, where drying occurs in the falling rate period with the constant drying rate period said to be negligible.

Methods of Drying

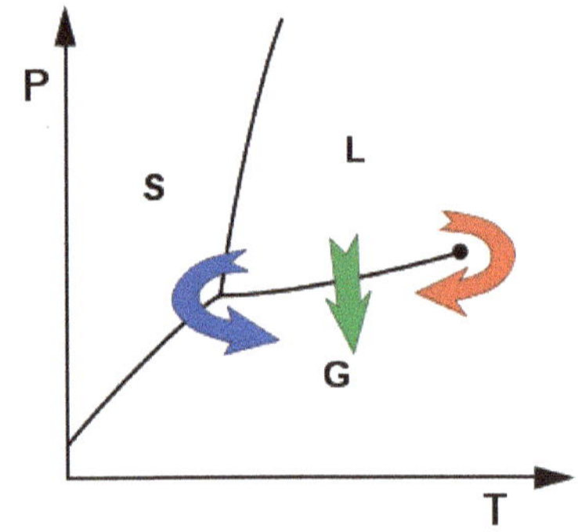

In a typical phase diagram, the boundary between gas and liquid runs from the triple point to the critical point. Regular drying is the green arrow, while supercritical drying is the red arrow and freeze drying is the blue.

The following are some general methods of drying:

- Application of hot air (convective or direct drying). Air heating increases the driving force for heat transfer and accelerates drying. It also reduces air relative humidity, further increasing the driving force for drying. In the falling rate period, as moisture content falls, the solids heat up and the higher temperatures speed up diffusion of water from the interior of the solid to the surface. However, product quality considerations limit the applicable rise to air temperature. Excessively hot air can almost completely dehydrate the solid surface, so that its pores shrink and almost close, leading to crust formation or "case hardening", which is usually undesirable. For instance in wood (timber) drying, air is heated (which speeds up drying) though some steam is also added to it (which hinders drying rate to a certain extent) in order to avoid excessive surface dehydration and product deformation owing to high moisture gradients across timber thickness. Spray drying belongs in this category.
- Indirect or contact drying (heating through a hot wall), as drum drying, vacuum drying. Again, higher wall temperatures will speed up drying but this is limited by product degradation or case-hardening. Drum drying belongs in this category.
- Dielectric drying (radiofrequency or microwaves being absorbed inside the material) is the focus of intense research nowadays. It may be used to assist air drying or vacuum drying. Researchers have found that microwave finish drying speeds up the otherwise very low drying rate at the end of the classical drying methods.
- Freeze drying or lyophilization is a drying method where the solvent is frozen prior to drying and is then sublimed, i.e., passed to the gas phase directly from the solid phase, below the melting point of the solvent. It is increasingly applied to dry foods, beyond its already classical pharmaceutical or medical applications. It keeps biological properties of proteins, and retains vitamins and bioactive compounds. Pressure can be reduced by a high vacuum pump (though freeze drying at atmospheric pressure is possible in dry air). If using a vacuum pump, the vapor produced by sublimation is removed from the system by converting it into ice in a condenser, operating at very low temperatures, outside the freeze drying chamber.
- Supercritical drying (superheated steam drying) involves steam drying of products containing water. This process is feasible because water in the product is boiled off, and joined with the drying medium, increasing its flow. It is usually employed in closed circuit and allows a proportion of latent heat to be recovered by recompression, a feature which is not possible with conventional air drying, for instance. The process has potential for use in foods if carried out at reduced pressure, to lower the boiling point.
- Natural air drying takes place when materials are dried with unheated forced air, taking advantage of its natural drying potential. The process is slow and weather-dependent, so a wise strategy "fan off-fan on" must be devised considering the following conditions: Air temperature, relative humidity and moisture content and temperature of the material being dried. Grains are increasingly dried with

this technique, and the total time (including fan off and on periods) may last from one week to various months, if a winter rest can be tolerated in cold areas.

Applications of Drying

Drying of fish in Lofoten in the production of stockfish

Food

Foods are dried to inhibit microbial development and quality decay. However, the extent of drying depends on product end-use. Cereals and oilseeds are dried after harvest to the moisture content that allows microbial stability during storage. Vegetables are blanched before drying to avoid rapid darkening, and drying is not only carried out to inhibit microbial growth, but also to avoid browning during storage. Concerning dried fruits, the reduction of moisture acts in combination with its acid and sugar contents to provide protection against microbial growth. Products such as milk powder must be dried to very low moisture contents in order to ensure flowability and avoid caking. This moisture is lower than that required to ensure inhibition to microbial development. Other products as crackers are dried beyond the microbial growth threshold to confer a crispy texture, which is liked by consumers.

Non-Food Products

Among non-food products, those that require considerable drying are wood (as part of Timber processing), paper, flax, and washing powder. The first two, owing to their organic origins, may develop mold if insufficiently dried. Another benefit of drying is a reduction in volume and weight.

Sludges and Fecal Materials from Sanitation Processes

In the area of sanitation, drying of sewage sludge from sewage treatment plants, fecal sludge or feces collected in urine-diverting dry toilets (UDDT) is a common method to achieve pathogen kill, as pathogens can only tolerate a certain dryness level. In addition, drying is required as a process step if the excreta based materials are meant to be incinerated.

ELECTROPHORESIS

Electrophoresis is the motion of dispersed particles relative to a fluid under the influence of a spatially uniform electric field. This electrokinetic phenomenon was observed for the first time in 1807 by Ferdinand Frederic Reuss (Moscow State University), who noticed that the application of a constant electric field caused clay particles dispersed in water to migrate. It is ultimately caused by the presence of a charged interface between the particle surface and the surrounding fluid. It is the basis for a number of analytical techniques used in chemistry for separating molecules by size, charge, or binding affinity.

Electrophoresis of positively charged particles (cations) is called cataphoresis, while electrophoresis of negatively charged particles (anions) is called anaphoresis. Electrophoresis is a technique used in laboratories in order to separate macromolecules based on size. The technique applies a negative charge so proteins move towards a positive charge. This is used for both DNA and RNA analysis. Polyacrylamide gel electrophoresis (PAGE) has a clearer resolution than agarose and is more suitable for quantitative analysis. In this technique DNA foot-printing can identify how proteins bind to DNA. It can be used to separate proteins by size, density and purity. It can also be used for plasmid analysis, which develops our understanding of bacteria becoming resistant to antibiotics.

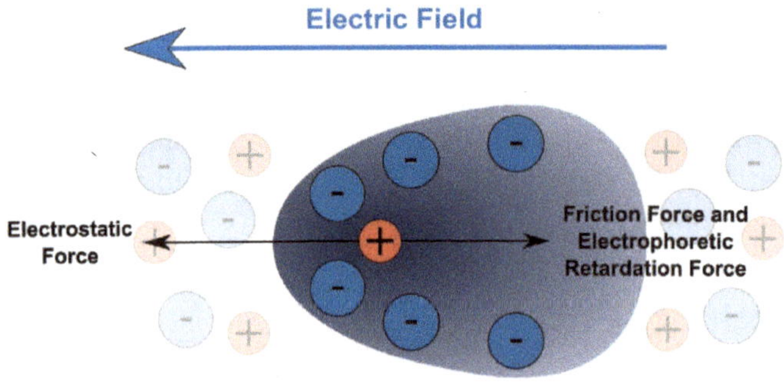

Illustration of electrophoresis

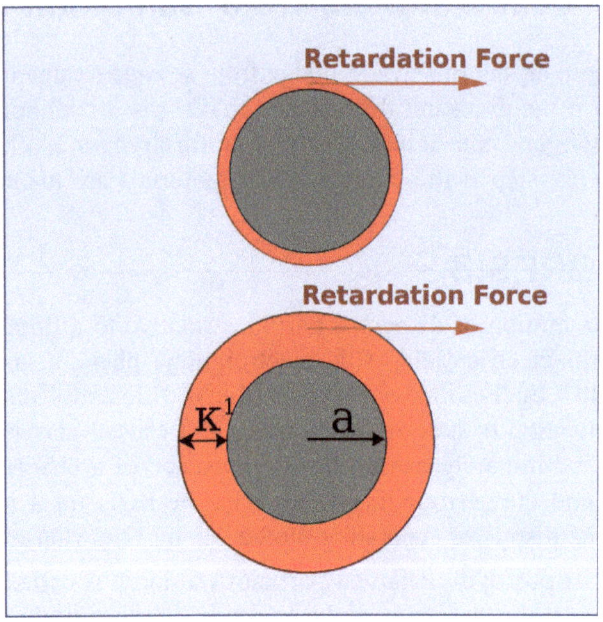

Illustration of electrophoresis retardation

Theory

Suspended particles have an electric surface charge, strongly affected by surface adsorbed species, on which an external electric field exerts an electrostatic Coulomb force. According to the double layer theory, all surface charges in fluids are screened by a diffuse layer of ions, which has the same absolute charge but opposite sign with respect to that of the surface charge. The electric field also exerts a force on the ions in the diffuse layer which has direction opposite to that acting on the surface charge. This latter force is not actually applied to the particle, but to the ions in the diffuse layer located at some distance from the particle surface, and part of it is transferred all the way to the particle surface through viscousstress. This part of the force is also called electrophoretic retardation force. When the electric field is applied and the charged particle to be analyzed is at steady movement through the diffuse layer, the total resulting force is zero :

$$F_{tot} = 0 = F_{el} + F_f + F_{ret}$$

Considering the drag on the moving particles due to the viscosity of the dispersant, in the case of low Reynolds number and moderate electric field strength E, the drift velocity of a dispersed particle v is simply proportional to the applied field, which leaves the electrophoretic mobility μ_e defined as:

$$\mu_e = \frac{v}{E}.$$

The most well known and widely used theory of electrophoresis was developed in 1903 by Smoluchowski:

$$\mu_e = \frac{\varepsilon_r \varepsilon_0 \zeta}{\eta},$$

where ε_r is the dielectric constant of the dispersion medium, ε_0 is the permittivity of free space ($C^2\ N^{-1}\ m^{-2}$), η is dynamic viscosity of the dispersion medium (Pa s), and ζ is zeta potential (i.e., the electrokinetic potential of the slipping plane in the double layer).

The Smoluchowski theory is very powerful because it works for dispersed particles of any shape at any concentration. It has limitations on its validity. It follows, for instance, because it does not include Debye length κ^{-1}. However, Debye length must be important for electrophoresis, as follows immediately from the Figure on the right. Increasing thickness of the double layer (DL) leads to removing the point of retardation force further from the particle surface. The thicker the DL, the smaller the retardation force must be.

Detailed theoretical analysis proved that the Smoluchowski theory is valid only for sufficiently thin DL, when particle radius a is much greater than the Debye length:

$$a\kappa \gg 1.$$

This model of "thin double layer" offers tremendous simplifications not only for electrophoresis theory but for many other electrokinetic theories. This model is valid for most aqueous systems, where the Debye length is usually only a few nanometers. It only breaks for nano-colloids in solution with ionic strength close to water.

The Smoluchowski theory also neglects the contributions from surface conductivity. This is expressed in modern theory as condition of small Dukhin number:

$$Du \ll 1$$

In the effort of expanding the range of validity of electrophoretic theories, the opposite asymptotic case was considered, when Debye length is larger than particle radius:

$$a\kappa < 1.$$

Under this condition of a "thick double layer", Hückel predicted the following relation for electrophoretic mobility:

$$\mu_e = \frac{2\varepsilon_r \varepsilon_0 \zeta}{3\eta}.$$

This model can be useful for some nanoparticles and non-polar fluids, where Debye length is much larger than in the usual cases.

There are several analytical theories that incorporate surface conductivity and eliminate the restriction of a small Dukhin number, pioneered by Overbeek. and Booth. Modern, rigorous theories valid for any Zeta potential and often any $a\kappa$ stem mostly from Dukhin–Semenikhin theory. In the **thin double layer** limit, these theories confirm the numerical solution to the problem provided by O'Brien and White.

MINERAL PROCESSING

In the field of extractive metallurgy, **mineral processing**, also known as **ore dressing**, is the process of separating commercially valuable minerals from their ores.

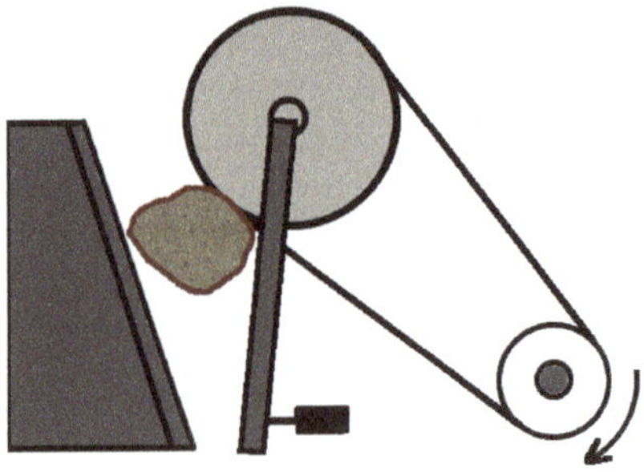

Crushing, a form of comminution, one of the unit operations of mineral processing

A set of Cornish stamps

Before the advent of heavy machinery the raw ore was broken up using hammers wielded by hand, a process called "spalling". Before long, mechanical means were found to achieve this. For instance, stamp mills were used in Samarkand as early as 973. They were also in use in medieval Persia. By the 11th century, stamp mills were in widespread use throughout the medieval Islamic world, from Islamic Spain and North Africa in the west to Central Asia in the east. A later example was the Cornish stamps, consisting of a series of iron hammers mounted in a vertical frame, raised by cams on the shaft of a waterwheel and falling onto the ore under gravity.

The simplest method of separating ore from gangue consists of picking out the individual crystals of each. This is a very tedious process, particularly when the individual particles are small. Another comparatively simple method relies on the various minerals having different densities, causing them to collect in different places: metallic minerals (being heavier) will drop out of suspension more quickly than lighter ones, which will be carried further by a stream of water. The process of panning and sifting for gold uses both of these methods. Various devices known as 'buddles' were used to take advantage of this property. Later, more advanced machines were used such as the Frue vanner, invented in 1874.

Other equipment used historically includes the hutch, a trough used with some ore-dressing machines and the keeve or kieve, a large tub used for differential settlement.

Unit Operations

Mineral processing can involve four general types of unit operation: *comminution* – particle size reduction; *sizing* – separation of particle sizes by screening or classification; *concentration* by taking advantage of physical and surface chemical properties; and *dewatering* – solid/liquid separation. In all of these processes, the most important considerations are the economics of the processes and this is dictated by the grade and recovery of the final product. To do this, the mineralogy of the ore needs to be considered as this dictates the amount of liberation required and the processes that can occur. The smaller the particles processes, the greater the theoretical grade and recovery of the final product, but this however is difficult to do with fine particles as they prevent certain concentration processes from occurring.

Comminution

Comminution is particle size reduction of materials. Comminution may be carried out on either dry materials or slurries. Crushing and grinding are the two primary comminution processes. Crushing is normally carried out on "run-of-mine" ore, while grinding (normally carried out after crushing) may be conducted on dry or slurried material. In comminution, the size reduction of particles is done by three types of forces: compression, impact and attrition. Compression and impact forces are extensively used in crushing operations while attrition is the dominant force in grinding. The primarily used equipment in crushing are-jaw crushers, gyratory crushers and cone crushers whereas rod mills and ball mills, closed circuited with a classifier unit, are generally employed for grinding purposes in a mineral processing plant. Crushing is a dry process whereas grinding is generally performed wet and hence is more energy intensive.

Sizing

Screening ore through a sieve, Fixed Nitrogen Research Laboratory, 1930

Sizer 2000 for screening coarse to small particles

Sizing is the general term for separation of particles according to their size.

The simplest sizing process is screening, or passing the particles to be sized through a screen or number of screens. Screening equipment can include grizzlies, bar screens, wedge wire screens, radial sieves, banana screens, multi-deck screens, vibratory screen, fine screens, flip flop screens and wire mesh screens. Screens can be static (typically the case for very coarse material), or they can incorporate mechanisms to shake or vibrate the screen. Some considerations in this process includes the screen material, the aperture size, shape and orientation, the amount of near sized particles, the addition of water, the amplitude and frequency of the vibrations, the angle of inclination, the presence of harmful materials, like steel and wood, and the size distribution of the particles.

Classification refers to sizing operations that exploit the differences in settling velocities exhibited by particles of different size. Classification equipment may include ore sorters, gas cyclones, hydrocyclones, rotating trommels, rake classifiers or fluidized classifiers.

An important factor in both comminution and sizing operations is the determination of the particle size distribution of the materials being processed, commonly referred to as particle size analysis. Many techniques for analyzing particle size are used, and the techniques include both off-line analyses which require that a sample of the material be taken for analysis and on-line techniques that allow for analysis of the material as it flows through the process.

Concentration

There are a number of ways to increase the concentration of the wanted minerals: in any particular case the method chosen will depend on the relative physical and surface chemical properties of the mineral and the gangue. Concentration is defined as the number of moles of a solute in a volume of the solution. In case of mineral processing concentration means the increase of the percentage of the valuable mineral in the concentrate.

Gravity Concentration

Gravity separation is the separation of two or more minerals of different specific gravity by their relative movement in response to the force of gravity and one or more other forces (such as centrifugal forces, magnetic forces, buoyant forces), one of which is resistance to motion (drag force) by a viscous medium such as heavy media, water or, less commonly, air.

Gravity separation is one of the oldest technique in mineral processing but has seen a decline in its use since the introduction of methods like flotation, classification, magnetic separation and leaching. Gravity separation dates back to at least 3000 BC when Egyptians used the technique for separation of gold.

It is necessary to determine the suitability of a gravity concentration process before it is employed for concentration of an ore. The *concentration criterion* is commonly used for this purpose, designated CC in the following equation (where SG represents specific gravity):

$$CC = \frac{SG(heavy\ mineral) - SG(fluid)}{SG(light\ mineral) - SG(fluid)}$$

- for CC > 2.5, suitable for separation of particles above 75 micron in size
- for 1.75 < CC < 2.5, suitable for separation of particles above 150 micron in size
- for 1.50 < CC < 1.75, suitable for separation of particles above 1.7 mm in size
- for 1.25 < CC < 1.50, suitable for separation of particles above 6.35 mm in size
- for CC < 1.25, not suitable for any size

Although concentration criteria is a useful rule of thumb when predicting amenability to gravity concentration, factors such as particle shape and relative concentration of heavy and light particles can dramatically affect separation efficiency in practice.

There are several methods that make use of the weight or density differences of particles:

- Heavy media or dense media separation (these include the Sepro Condor DMS, baths, drums, larcodems, dyana whirlpool separators, and dense medium cyclones)
- Shaking tables, such as the Wilfley table
- Spiral separators
- Reflux Classifier
- Jig concentrators are continuous processing gravity concentration devices using a pulsating fluidized bed.(RMS-Ross Corp. Circular Jig Plants)
- Centrifugal bowl concentrators, such as the Knelson concentrator and Falcon Concentrator
- Multi gravity separators (Falcon Concentrator, Knelson, Mozley (Multi or Enhanced Gravity Separator now manufactured by Gravity Mining Ltd, England. www.gravitymining.com), Salter Cyclones (Multi-Gravity Separator) and the Kelsey Jig)
- Inline pressure Jigs
- Reichert Cones
- Sluices
- Elutriators

These processes can be classified as either density separation or gravity (weight) separation.

In dense media separation a media is created with a density in between the density of the ore and gangue particles. When subjected to this media particles either float or sink depending on their density relative to the media. In this way the separation takes place purely on density differences and does not, in principle, relay on any other factors such as particle weight or shape. In practice, particle size and shape can affect separation efficiency. Dense medium separation can be performed using a variety of mediums. These include, organic liquids, aqueous solutions or suspensions of very fine particles in water or air. The organic liquids are typically not used due to their toxicity, difficulties in handling and relative cost. Industrially, the most common dense media is a suspension of fine magnetite and/or ferrosilicon particles. An aqueous solution as a dense medium is used in coal processing in the form of a belknap wash and suspensions in air are used in water-deficient areas, like areas of China, where sand is used to separate coal from the gangue minerals.

Gravity separation is also called relative gravity separation as it separates particles due to their relative response to a driving force. This is controlled by factors such as particle weight, size and shape. These processes can also be classified into multi-G and single G processes. The difference is the magnitude of the driving force for the separation. Multi-G processes allow the separation of very fine particles to occur (in the range of 5 to 50 micron) by increasing the driving force of separation in order to increase the rate at which particles separate. In general, single G process are only capable of processing particles that are greater than approximately 80 micron in diameter.

Of the gravity separation processes, the spiral concentrators and circular jigs are two of the most economical due to their simplicity and use of space. They operate by flowing film separation and can either use washwater or be washwater-less. The washwater spirals separate particles more easily but can have issues with entrainment of gangue with the concentrate produced.

Froth Flotation

Froth flotation cells used to concentrate copper and nickel sulfide minerals, Falconbridge, Ontario.

Froth flotation is an important concentration process. This process can be used to separate any two different particles and operated by the surface chemistry of the particles. In flotation, bubbles are introduced into a pulp and the bubbles rise through the pulp. In the process, hydrophobic particles become bound to the surface of the bubbles. The driving force for this attachment is the change in the surface free energy when the attachment occurs. These bubbles rise through the slurry and are collected from the surface. To enable these particles to attach, careful consideration of the chemistry of the pulp needs to be made. These considerations include the pH, Eh and the presence of flotation reagents. The pH is important as it changes the charge of the particles surface and the Eh affects the chemisorption of collectors on the surface of the particles.

The addition of flotation reagents also affects the operation of these processes. The most important chemical that is added is the collector. This chemical binds to the surface of the particles as it is a surfactant. The main considerations in this chemical is the nature of the head group and the size of the hydrocarbon chain. The hydrocarbon tail needs to be short to maximize the selectivity of the desired mineral and the headgroup dictates which minerals it attaches to.

The frothers are another important chemical addition to the pulp as it enables stable bubbles to be formed. This is important as if the bubble coalesce, minerals fall off their surface. The bubbles however should not be too stable as this prevents easy transportation and dewatering of the concentrate formed. The mechanism of these frothers is not completely known and further research into their mechanisms is being performed.

Depressants and activators are used to selectively separate one mineral from another. Depressants inhibit the flotation of one mineral or minerals while activators enable the flotation of others. Examples of these include CN^-, used to depress all sulfides but galena and this depressant is believed to operate by changing the solubility of chemisorbed and physisorbed collectors on sulfides. This theory originates from Russia. An example of an activator is Cu^{2+} ions, used for the flotation of sphalerite.

There are a number of cells able to be used for the flotation of minerals. these include flotation columns and mechanical flotation cells. The flotation columns are used for finer minerals and they typically have a higher grade and lower recovery of minerals than mechanical flotation cells. The cells in use at the moment can exceed 300 m^3. This is done as they are cheaper per unit volume than smaller cells, but they are not able to be controlled as easily as smaller cells.

This process was invented in the 19th century in Australia. It was used to recover a sphalerite concentrate from tailings, produced using gravity concentration. Further improvements have come from Australia in the form of the Jameson Cell, developed at the University of Newcastle, Australia. This operated by the use of a plunging jet that generates fine bubbles. These fine bubbles have a higher kinetic energy and as such they can be used for the flotation of fine grained minerals, such as those produced by the isamill.

Electrostatic Separation

There are two main types of electrostatic separators. These work in similar ways, but the forces applied to the particles are different and these forces are gravity and electrostatic attraction. The two types are electrodynamic separators (or high tension rollers) or electrostatic separators. In high tension rollers, particles are charged by a corona discharge. This charges the particles that subsequently travel on a drum. The conducting particles lose their charge to the drum and are removed from the drum with centripetal acceleration. Electrostatic plate separators work by passing a stream of particles past a charged anode. The conductors lose electrons to the plate and are pulled away from the other particles due to the induced attraction to the anode. These separators are used for particles between 75 and 250 micron and for efficient separation to occur, the particles need to be dry, have a close size distribution and uniform in shape. Of these considerations, one of the most important is the water content of the particles. This is important as a layer of moisture on the particles will render the non-conductors as conductors as the layer of the water is conductive.

Electrostatic plate separators are usually used for streams that have small conductors and coarse non-conductors. The high tension rollers are usually used for streams that have coarse conductors and fine non-conductors.

These separators are commonly used for separating mineral sands, an example of one of these mineral processing plants is the CRL processing plant at Pinkenba in Brisbane Queensland. In this plant, zircon, rutile and ilmenite are separated from the silica gangue. In this plant, the separation is performed in a number of stages with roughers, cleaners, scavengers and recleaners.

Magnetic Separation

Magnetic separation is a process in which magnetically susceptible material is extracted from a mixture using a magnetic force. This separation technique can be useful in mining iron as it is attracted to a magnet. In mines where wolframite was mixed with cassiterite, such as South Crofty and East Pool mine in Cornwall or with bismuth such as at the Shepherd and Murphy mine in Moina, Tasmania, magnetic separation was used to separate the ores. At these mines a device called a Wetherill's Magnetic Separator (invented by John Price Wetherill, 1844–1906) was used. In this machine the raw ore, after calcination was fed onto a moving belt which passed underneath two pairs of electromagnets under which further belts ran at right angles to the feed belt. The first pair of electromagnets was weakly magnetised and served to draw off any iron ore present. The second pair were strongly magnetised and attracted the wolframite, which is weakly magnetic. These machines were capable of treating 10 tons of ore a day.This process of separating magnetic substances from the non-magnetic substances in a mixture with the help of a magnet is called magnetic separation..

This process operates by moving particles in a magnetic field. The force experienced in the magnetic field is given by the equation $f=m/k.H.dh/dx$. with k=magnetic susceptibility, H-magnetic field strength, and dh/dx being the magnetic field gradient. As seen in this equation, the separation can be driven in two ways, either through a gradient in a magnetic field or the strength of a magnetic field. The different driving forces are used in the different concentrators. These can be either with water or without. Like the spirals, washwater aids in the separation of the particles while increases the entrainment of the gangue in the concentrate.

Automated Ore Sorting

Modern, automated sorting applies optical sensors (visible spectrum, near infrared, X-ray, ultraviolet), that can be coupled with electrical conductivity and magnetic susceptibility sensors, to control the mechanical separation of ore into two or more categories on an individual rock by rock basis. Also new sensors have been developed which exploit material properties such as electrical conductivity, magnetization, molecular structure and thermal conductivity. Sensor based sorting has found application in the processing of nickel, gold, copper, coal and diamonds.

Dewatering

Dewatering is an important process in mineral processing. The purpose of dewatering is to remove water absorbed by the particles which increases the pulp density. This is done for a number of reasons, specifically, to enable ore handling and concentrates to be transported easily, allow further processing to occur and to dispose of the gangue. The water extracted from the ore by dewatering is recirculated for plant operations after being sent to a water treatment plant. The main processes that are used in dewatering include dewatering screens such as Sepro-Sizetec Screens, sedimentation, filtering, and thermal drying. These processes increase in difficulty and cost as the particle size decreases.

Dewatering screens operate by passing particles over a screen. The particles pass over the screen while the water passes through the apertures in the screen. This process is only viable for coarse ores that have a close size distribution as the apertures can allow small particles to pass through.

Sedimentation operates by passing water into a large thickener or clarifier. In these devices, the particles settle out of the slurry under the effects of gravity, or centripetal forces. These are limited by the surface chemistry of the particles and the size of the particles. To aid in the sedimentation process, flocculants and coagulants are added to reduce the repulsive forces between the particles. This repulsive force is due to the double layer formed on the surface of the particles. The flocculants work by binding multiple particles together while the coagulants work by reducing the thickness of the charged layer on the outside of the particle. After thickening, slurry is often stored in ponds or impoundments. Alternatively, it can pumped into a belt press or membrane filter press to recycle process water and create stackable, dry filter cake, or "tailings".

Thermal drying is usually used for fine particles and to remove low water content in the particles. Some common processes include rotary dryers, fluidised beds, spray driers, hearth dryers and rotary tray dryers. This process is usually expensive to operate due to the fuel requirement of the dryers.

Other Processes

Many mechanical plants also incorporate hydrometallurgical or pyrometallurgical processes as part of an extractive metallurgical operation. Geometallurgy is a branch of extractive metallurgy that combines mineral processing with the geologic sciences. We study the oil agglomeration

A number of auxiliary materials handling operations are also considered a branch of mineral processing such as storage (as in bin design), conveying, sampling, weighing, slurry transport, and pneumatic transport.

The efficiency and efficacy of many processing techniques are influenced by upstream activities such as mining method and blending.

ELUTRIATION

Elutriation is a process for separating particles based on their size, shape and density, using a stream of gas or liquid flowing in a direction usually opposite to the direction of sedimentation. This method is mainly used for particles smaller than 1 µm. The smaller or lighter particles rise to the top (overflow) because their terminal sedimentation velocities are lower than the velocity of the rising fluid. The terminal velocity of any particle in any medium can be calculated using Stokes› Law if the particle›s Reynolds number is below 0.2. Counterflow centrifugation elutriation is a related technique to separate cells.

Air Classification

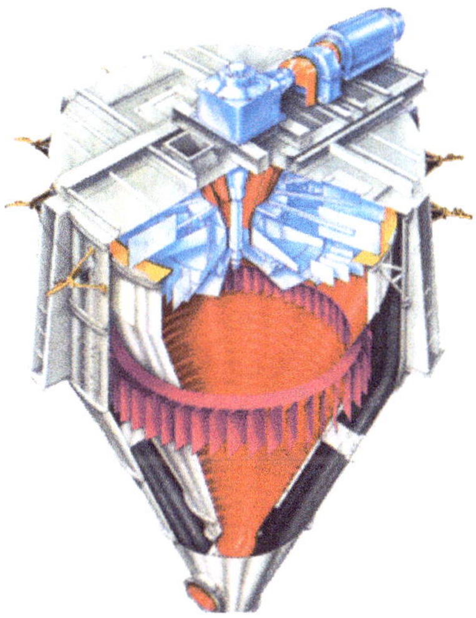

Air classifier

An air elutriator is a simple device which can separate particles into two or more groups.

Material may be separated by means of an elutriator, which consists of a vertical tube up which fluid is passed at a controlled velocity. When the particles are introduced, often through a side tube, the smaller particles are carried over in the fluid stream while the larger particles settle against the upward current. If one starts with low flow rates, small less dense particle attain their terminal velocities, and flow with the stream. The particle from the stream is collected in overflow and hence will be separated from the feed. Flow rates can be increased to separate higher size ranges. Further size fractions may be collected if the overflow from the first tube is passed vertically upwards through a second tube of greater cross-section, and any number of such tubes can be arranged in series.

It is used in mineral processing for size classification. The elutriation dust value is a usual measure for quantification of dust, generated by testing wherein mechanical forces such as vibration are applied to granules of e.g. a detergent agent.

Elutriation is a common method used by biologists to sample meiofauna. The sediment sample is constantly agitated by a flow of filtered water from below, the action of which dislodges interstitial organisms embedded between sediment grains. A very fine filter at the top captures these organisms from the overflow.

EVAPORATION

Evaporation is a type of vaporization of a liquid that occurs from the surface of a liquid into a gaseous phase that is not saturated with the evaporating substance. The other type of vaporization is boiling, which is characterized by bubbles of saturated vapor forming in the liquid phase. Steam produced in a boiler is another example of evaporation occurring in a saturated vapor phase. Evaporation that occurs directly from the solid phase below the melting point, as commonly observed with ice at or below freezing or moth crystals (napthalene or paradichlorobenzene), is called sublimation.

On average, a fraction of the molecules in a glass of water have enough heat energy to escape from the liquid. The reverse also happens — water molecules from the air enter the water in the glass — but as long as the relative humidity of the air in contact is less than 100% (i.e., saturation), the net transfer of water molecules will be to the air. The water in the glass will be cooled by the evaporation until an equilibrium is reached where the air supplies the amount of heat removed by the evaporating water. In an enclosed environment the water would evaporate until the air is saturated.

With sufficient temperature, the liquid would turn into vapor quickly (see boiling point). When the molecules collide, they transfer energy to each other in varying degrees, based on how they collide. Sometimes the transfer is so one-sided for a molecule near the surface that it ends up with enough energy to escape and enter the surrounding air.

Evaporation is an essential part of the water cycle. The sun (solar energy) drives evaporation of water from oceans, lakes, moisture in the soil, and other sources of water. In hydrology, evaporation and transpiration (which involves evaporation within plant stomata) are collectively termed evapotranspiration. Evaporation of water occurs when the surface of the liquid is exposed, allowing molecules to escape and form water vapor; this vapor can then rise up and form clouds.

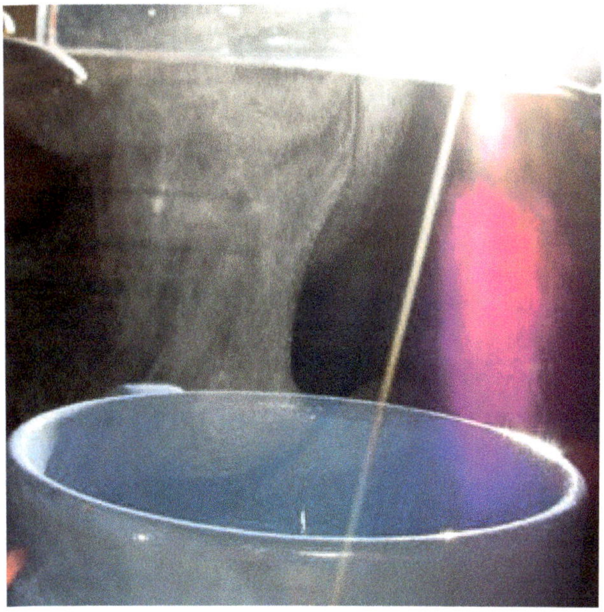

Aerosol of microscopic water droplets suspended in the air above a hot tea cup after that water vapor has sufficiently cooled and condensed. Water vapor is an invisible gas, but the clouds of condensed water droplets refract and disperse the sun light and so are visible.

Theory

For molecules of a liquid to evaporate, they must be located near the surface, they have to be moving in the proper direction, and have sufficient kinetic energy to overcome liquid-phase intermolecular forces. When only a small proportion of the molecules meet these criteria, the rate of evaporation is low. Since the kinetic energy of a molecule is proportional to its temperature, evaporation proceeds more quickly at higher temperatures. As the faster-moving molecules escape, the remaining molecules have lower average kinetic energy, and the temperature of the liquid decreases. This phenomenon is also called evaporative cooling. This is why evaporating sweat cools the human body. Evaporation also tends to proceed more quickly with higher flow rates between the gaseous and liquid phase and in liquids with higher vapor pressure. For example, laundry on a clothes line will dry (by evaporation) more rapidly on a windy day than on a still day. Three key parts to evaporation are heat, atmospheric pressure (determines the percent humidity), and air movement.

On a molecular level, there is no strict boundary between the liquid state and the vapor state. Instead, there is a Knudsen layer, where the phase is undetermined. Because this layer is only a few molecules thick, at a macroscopic scale a clear phase transition interface cannot be seen.

Liquids that do not evaporate visibly at a given temperature in a given gas (e.g., cooking oil at room temperature) have molecules that do not tend to transfer energy to each other in a pattern sufficient to frequently give a molecule the heat energy necessary to turn into vapor. However, these liquids *are* evaporating. It is just that the process is much slower and thus significantly less visible.

Evaporative Equilibrium

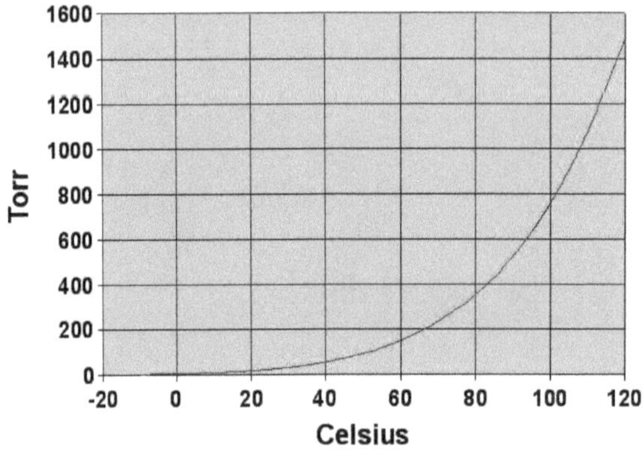

Vapor pressure of water vs. temperature. 760 Torr = 1 atm.

If evaporation takes place in an enclosed area, the escaping molecules accumulate as a vapor above the liquid. Many of the molecules return to the liquid, with returning molecules becoming more frequent as the density and pressure of the vapor increases. When the process of escape and return reaches an equilibrium, the vapor is said to be saturated, and no further change in either vapor pressure and density or liquid temperature will occur. For a system consisting of vapor and liquid of a pure substance, this equilibrium state is directly related to the vapor pressure of the substance, as given by the Clausius–Clapeyron relation:

$$\ln\left(\frac{P_2}{P_1}\right) = -\frac{\Delta H_{vap}}{R}\left(\frac{1}{T_2} - \frac{1}{T_1}\right)$$

where P_1, P_2 are the vapor pressures at temperatures T_1, T_2 respectively, ΔH_{vap} is the enthalpy of vaporization, and R is the universal gas constant. The rate of evaporation in an open system is related to the vapor pressure found in a closed system. If a liquid is heated, when the vapor pressure reaches the ambient pressure the liquid will boil.

The ability for a molecule of a liquid to evaporate is based largely on the amount of kinetic energy an individual particle may possess. Even at lower temperatures, individual molecules of a liquid can evaporate if they have more than the minimum amount of kinetic energy required for vaporization.

Factors Influencing the Rate of Evaporation

Note: Air used here is a common example; however, the vapor phase can be other gases.

Concentration of the substance evaporating in the air

If the air already has a high concentration of the substance evaporating, then the given substance will evaporate more slowly.

Concentration of other substances in the air

If the air is already saturated with other substances, it can have a lower capacity for the substance evaporating.

Flow rate of air

This is in part related to the concentration points above. If "fresh" air (i.e., air which is neither already saturated with the substance nor with other substances) is moving over the substance all the time, then the concentration of the substance in the air is less likely to go up with time, thus encouraging faster evaporation. This is the result of the boundary layer at the evaporation surface decreasing with flow velocity, decreasing the diffusion distance in the stagnant layer.

Inter-molecular forces

The stronger the forces keeping the molecules together in the liquid state, the more energy one must get to escape. This is characterized by the enthalpy of vaporization.

Pressure

Evaporation happens faster if there is less exertion on the surface keeping the molecules from launching themselves.

Surface area

A substance that has a larger surface area will evaporate faster, as there are more surface molecules per unit of volume that are potentially able to escape.

Temperature of the substance

the higher the temperature of the substance the greater the kinetic energy of the molecules at its surface and therefore the faster the rate of their evaporation.

In the US, the National Weather Service measures the actual rate of evaporation from a standardized "pan" open water surface outdoors, at various locations nationwide. Others do likewise around the world. The US data is collected and compiled into an annual evaporation map. The measurements range from under 30 to over 120 inches (3,000 mm) per year.

Thermodynamics

Evaporation is an endothermic process, in that heat is absorbed during evaporation.

Applications

- Industrial applications include many printing and coating processes; recovering salts from solutions; and drying a variety of materials such as lumber, paper, cloth and chemicals.
- The use of evaporation to dry or concentrate samples is a common preparatory step for many laboratory analyses such as spectroscopy and chromatography. Systems used for this purpose include rotary evaporators and centrifugal evaporators.
- When clothes are hung on a laundry line, even though the ambient temperature is below the boiling point of water, water evaporates. This is accelerated by factors such as low humidity, heat (from the sun), and wind. In a clothes dryer, hot air is blown through the clothes, allowing water to evaporate very rapidly.
- The Matki/Matka, a traditional Indian porous clay container used for storing and cooling water and other liquids.
- The botijo, a traditional Spanish porous clay container designed to cool the contained water by evaporation.
- Evaporative coolers, which can significantly cool a building by simply blowing dry air over a filter saturated with water.

Combustion Vaporization

Fuel droplets vaporize as they receive heat by mixing with the hot gases in the combustion chamber. Heat (energy) can also be received by radiation from any hot refractory wall of the combustion chamber.

Pre-Combustion Vaporization

Internal combustion engines rely upon the vaporization of the fuel in the cylinders to form a fuel/air mixture in order to burn well. The chemically correct air/fuel mixture for total burning of gasoline has been determined to be 15 parts air to one part gasoline or 15/1 by weight. Changing this to a volume ratio yields 8000 parts air to one part gasoline or 8,000/1 by volume.

Film Deposition

Thin films may be deposited by evaporating a substance and condensing it onto a substrate, or by dissolving the substance in a solvent, spreading the resulting solution thinly over a substrate, and evaporating the solvent. The Hertz–Knudsen equation is often used to estimate the rate of evaporation in these instances.

REFERENCES

1. "drying - definition of drying by the Free Online Dictionary, Thesaurus and Encyclopedia". Farlex. Retrieved 23 April 2011.
2. "Modeling the thin-layer drying of fruits and vegetables: A review". Wiley-Blackwell. 2016-02-04. Retrieved 2016-02-08.
3. Lyklema, J. (1995). Fundamentals of Interface and Colloid Science. vol. 2. p. 3.208.
4. Hunter, R.J. (1989). Foundations of Colloid Science. Oxford University Press.
5. Dukhin, S.S.; B.V. Derjaguin (1974). Electrokinetic Phenomena. J. Willey and Sons.
6. Russel, W.B.; D.A. Saville; W.R. Schowalter (1989). Colloidal Dispersions. Cambridge University Press.
7. Kruyt, H.R. (1952). Colloid Science. Volume 1, Irreversible systems. Elsevier.
8. Dukhin, A.S.; P.J. Goetz (2002). Ultrasound for characterizing colloids. Elsevier.
9. Reuss, F.F. (1809). "Sur un nouvel effet de l'électricité galvanique". Mémoires de la Société Impériale des Naturalistes de Moscou. 2: 327–337.
10. Hanaor, D.A.H.; Michelazzi, M.; Leonelli, C.; Sorrell, C.C. (2012). "The effects of carboxylic acids on the aqueous dispersion and electrophoretic deposition of ZrO2". Journal of the European Ceramic Society. 32 (1): 235–244. doi:10.1016/j.jeurceramsoc.2011.08.015.
11. von Smoluchowski, M. (1903). "Contribution à la théorie de l'endosmose électrique et de quelques phénomènes corrélatifs. Bull. Int. Acad. Sci. Cracovie. 184.
12. Hückel, E. (1924). Die kataphorese der kugel. Physik. Z. 25: 204.
13. Overbeek, J.Th.G (1943). Theory of electrophoresis — The relaxation effect. Koll. Bith.: 287.
14. Booth, F. (1948). Theory of Electrokinetic Effects. Naturc. 161 (4081): 83–86. Bibcode:1948Natur.161...83B. doi:10.1038/161083a0. PMID 18898334.
15. Dukhin, S.S.; N.M. Semenikhin (1970). Koll. Zhur. 32: 366. Missing or empty |title= (help)
16. O›Brien, R.W.; L.R. White (1978). Electrophoretic mobility of a spherical colloidal particle. J. Chem. Soc. Faraday Trans. 2 (74): 1607. doi:10.1039/F29787401607.
17. Adam Robert Lucas (2005), Industrial Milling in the Ancient and Medieval Worlds: A Survey of the Evidence for an Industrial Revolution in Medieval Europe, Technology and Culture 46 (1): 1-30 10-1 & 27
18. Run-of-mine: The raw mined material as it is delivered prior to treatment of any sort. Dictionary of Mining, Mineral, and Related Terms. Hacettepe University - Department of Mining Engineering. Retrieved 2010-08-07.
19. Grizzly: a grid of iron bars that allows ore of the correct size to travel down

the ore pass to the bottom of the mine, ready for hoisting to the surface. An active, articulating grizzly that is able to roll, scrub, clean and discharge oversize rock and boulder of up to 4 foot (1220 mm minus)diameter, while recovering all the 2 inch minus (51 mm minus) slurry material for further screening, separation and recovery of target metals/minerals is the DEROCKER system (RMS-Ross Corporation) Geevor Tin Mine: Grizzly men. Geevor Tin Mine Museum. Retrieved 2010-08-07.

20. Mill Machines: The Wilfley Table. Copper Country Explorer. Retrieved 2010-08-07.
21. Tons Per Hour Product Guide 2016 (PDF).
22. Preparation of salty coals with oil-agglomeration technology / V. S. Biletskyi, A. Kheloufi, P. V. Sergeev // 9th International conference on coal science (ICCS'97), 7-12 September, 1997, Essen, Germany. V. 1. P.535-538.
23. Investigation of laws of selective flocculation of coals with syn-thetic latexes / P. V. Sergeev, V. S. Biletskyy // ICCS'97. 7-12 September, 1997, Essen, Germany. V. 1. p. 503-506.
24. C.-W. Fan, R. Markuszewski, and T. D. Wheelock, Oil Agglomeration of Coal in Salt Solutions: Effects of Hydrophobicity and Other Parameters on Coal Recovery
25. Beletskyi V., Shendrik T. Ennobling of salty coals by means of oil agglomeration Technical and Geoinformational Systems in Mining. Proceedings of the School of underground mining, Dnipropetrovsk/Yalta, 2-8 October 2011. / CRC Press Taylor & Francis Group, London, UK. A Balkema Book. 2011. p. 135-140.
26. Whitacre, J., Iorio, A., Schellenberg, S. Coal Blending: Business Value, Analysis, and Optimization

CHAPTER 6

Extraction Chemistry

Chapter Contents

Extractions are a way to separate a desired substance when it is mixed with others. The mixture is brought into contact with a solvent in which the substance of interest is soluble, but the other substances present are insoluble.

INTRODUCTION

Extraction in chemistry is a separation process consisting in the separation of a substance from a matrix. It includes Liquid-liquid extraction, and Solid phase extraction.

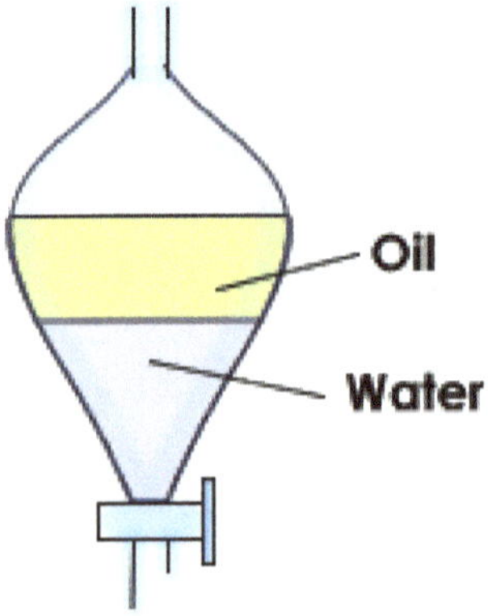

Laboratory-scale liquid-liquid extraction. Schematic of a separatory funnel used for laboratory scale (<2L combined volume) extractions of 2 immiscible liquid phases, where "oil" may be any phase less dense than water (typically an organic solvent) and "water" may be any less dense liquid phase, typically an aqueous phase.

Laboratory-scale liquid-liquid extraction. Photograph of a separatory funnel in a laboratory scale extraction of 2 immiscible liquids, see above; liquids are a diethyl ether upper phase, and a lower aqueous phase.

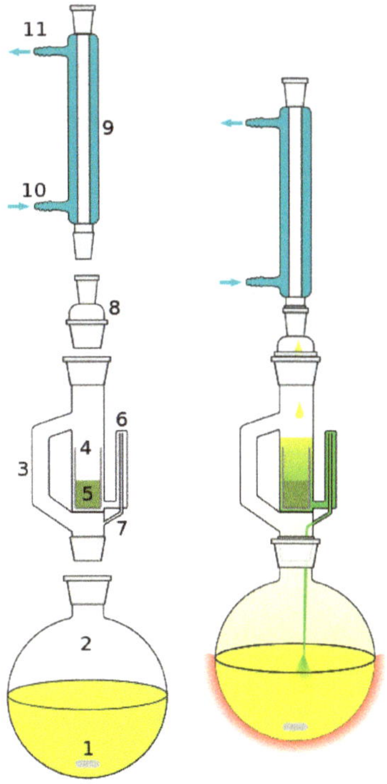

Soxhlet extractor

Types of Extraction

There exist several types of extraction, including: *liquid-liquid extraction*, *solid-phase extraction*, *solid-phase microextraction*, *Soxhlet extraction*, *fizzy extraction*.

Components of an Extractive Process

Extractions often use two immiscible phases to separate a solute from one phase into the other. Typical lab extractions are of organic compounds out of an aqueous phase and into an organic phase. Common extractants are arranged from ethyl acetate to water (ethyl acetate < acetone < ethanol < methanol < acetone:water (7:3) < ethanol:water (8:2) < methanol:water (8:2) < water) in increasing order of polarity according to the Hildebrand solubility parameter. The extract can be put back to dried form using a centrifugal evaporator or a freeze-drier.

Theory of an Extraction between two Phases

The distribution of a solute between two phases is an equilibrium condition described by partition theory. This is based on exactly how the analyte move from the water into an organic layer

Modern Variations

Techniques include supercritical carbon dioxide extraction, ultrasonic extraction, heat reflux extraction, microwave-assisted extraction, instant controlled pressure drop extraction (DIC), and perstraction.

Applications

Boiling tea leaves in water extracts the tannins, theobromine, and caffeine out of the leaves and into the water. Solid-liquid extractions at laboratory scales can use Soxhlet extractors (such as oil from olive cakesee at right).

LEACHING (CHEMISTRY)

Leaching is the process of extracting substances from a solid by dissolving them in a liquid, either in nature or through an industrial process. In the chemical processing industry, leaching has a variety of commercial applications, including separation of metal from ore using acid, and sugar from beets using hot water.

Another term for this is lixiviation, or the extraction of a soluble particle from its constituent parts.

In a typical leaching operation, the solid mixture to be separated consists of particles, inert insoluble carrier A and solute B. The solvent, C, is added to the mixture to selectively dissolve B. The overflow from the stage is free of solids and consists of only solvent C and dissolved B. The underflow consists of slurry of liquid of similar composition in the liquid overflow and solid carrier A. In an ideal leaching equilibrium stage, all the solute is dissolved by the solvent; none of the carrier is dissolved. The mass ratio of the solid to liquid in the underflow is dependent on the type of equipment used and properties of the two phases.

Leaching is the process by which inorganic, organic contaminants or radionuclides are released from the solid phase into the water phase under the influence of mineral dissolution, desorption, complexation processes as affected by pH, redox, dissolved organic matter and (micro) biological activity. The process itself is universal, as any material exposed to contact with water will leach components from its surface or its interior depending on the porosity of the material considered.

One such reaction is:

$$Ag_2S + 4NaCN \rightarrow 2Na[Ag(CN)_2] + Na_2S$$

Leaching Processes for Biological Substances

Many biological organic and inorganic substances occur in a mixture of different components in a solid. In order to separate the desired solute constituent or remove an undesirable solute component from the solid phase, the solid is brought into contact with a liquid. The solid and liquid are in contact and the solute or solutes can diffuse from the solid into the solvent, resulting in separation of the components originally in the solid. This separation process is called *liquid-solid leaching* or simply *leaching*. Because in leaching the solute is being extracted from the solid this is also called *extraction*. In leaching, when an undesirable component is removed from a solid with water, the process is called *washing*.

In the biological and food processing industries, many products are separated from their original natural structure by liquid-solid leaching. An important process for example is the leaching of sugar from sugar beets with hot water. In the production of vegetable oils, organic solvents such as hexane, acetone, and/or ether are used to extract oil from nuts, beans and seeds.

In the pharmaceutical industry, many different pharmaceutical products are obtained by leaching plant roots, leaves, and stems.

Leaching Processes for Inorganic and Organic Materials

Leaching is extensively used in metal processing industries. The useful metal may occur in mixtures with very large amounts of undesirable constituents, and leaching is used to remove the metals as soluble salts. The use of acids is prevalent in the metal processing industry. Sulfates are normally used to remove metals from the solid phase. These processes result in harmful sulfate-rich environmental byproducts. Because acids solubilize metals, large-scale release of heavy-metal-rich acidic mine drainage (AMD) can thereby occur.

Shrinking-Core Model

As mentioned above when leaching a solid with a liquid the desired solid goes to the liquid phase while undesired solid remains. The removal of the solid as the liquid dissolves into the particle leads to a diameter of unleached core that shrinks with time. A mathematical model can be used for this process, derived using Fick's law of diffusion taking pore diffusion as rate limiting step:

$$t = \frac{\rho_B r_s^2}{6 D_e b M_B c_{A_b}} \left[1 - 3\left(\frac{r_c}{r_s}\right)^2 + 2\left(\frac{r_c}{r_s}\right)^3 \right]$$

Environmentally Friendly Leaching

Some recent work has been done to see if organic acids can be used to leach lithium and cobalt from spent batteries with some success. Experiments performed with varying temperatures and concentrations of malic acid show that the optimal conditions are 2.0 m/L of organic acid at a temperature of 90 °C. The reaction had an overall efficiency exceeding 90% with no harmful byproducts.

4 $LiCoO_2$(solid) + 12 $C_4H_6O_5$(liquid) → 4 $LiC_4H_5O_5$(liquid) + 4 $Co(C_4H_6O_5)_2$(liquid) + 6 H_2O(liquid) + O_2(gas)

The same analysis with citric acid showed similar results with an optimal temperature and concentration of 90 °C and 1.5 molar solution of citric acid.

LIQUID–LIQUID EXTRACTION

Liquid–liquid extraction (LLE) consists in transferring one (or more) solute(s) contained in a feed solution to another immiscible liquid (solvent). The solvent that is enriched in solute(s) is called extract. The feed solution that is depleted in solute(s) is called raffinate.

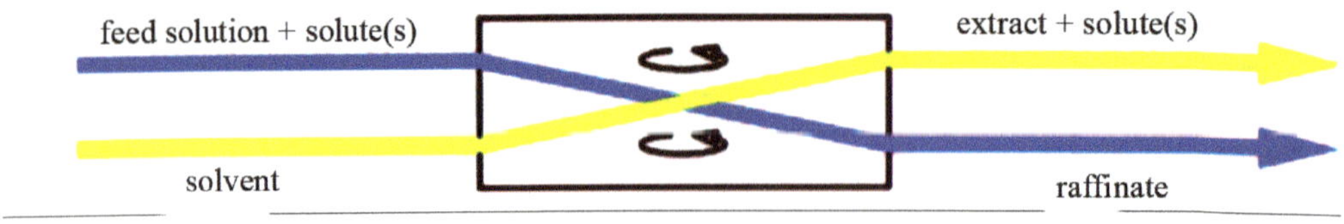

Solvent extraction.

Liquid–liquid extraction also known as **solvent extraction** and **partitioning**, is a method to separate compounds based on their relative solubilities in two different immiscible liquids, usually water and an organic solvent. It is an extraction of a substance from one liquid into another liquid phase. Liquid–liquid extraction is a basic technique in chemical laboratories, where it is performed using a variety of apparatus, from separatory funnels to countercurrent distribution equipment. This type of process is commonly performed after a chemical reaction as part of the work-up, often including an acidic work up.

The term *partitioning* is commonly used to refer to the underlying chemical and physical processes involved in *liquid–liquid extraction*, but on another reading may be fully synonymous with it. The term *solvent extraction* can also refer to the separation of a substance from a mixture by preferentially dissolving that substance in a suitable solvent. In that case, a soluble compound is separated from an insoluble compound or a complex matrix.

Solvent extraction is used in nuclear reprocessing, ore processing, the production of fine organic compounds, the processing of perfumes, the production of vegetable oils and biodiesel, and other industries. It is among the most common initial separation techniques, though some difficulties result in extracting out closely related functional groups.

Liquid–liquid extraction is possible in non-aqueous systems: In a system consisting of a molten metal in contact with molten salts, metals can be extracted from one phase to the other. This is related to a mercury electrode where a metal can be reduced, the metal will often then dissolve in the mercury to form an amalgam that modifies its electrochemistry greatly. For example, it is possible for sodium cations to be reduced at a mercury cathode to form sodium amalgam, while at an inert electrode (such as platinum) the sodium cations are not reduced. Instead, water is reduced to hydrogen. A detergent or fine solid can be used to stabilize an emulsion, or third phase.

Measures of Effectiveness

Distribution Ratio

In solvent extraction, a distribution ratio is often quoted as a measure of how well-extracted a species is. The distribution ratio (*Kd*) is equal to the concentration of a solute in the organic phase divided by its concentration in the aqueous phase. Depending on the system, the distribution ratio can be a function of temperature, the concentration of chemical species in the system, and a large number of other parameters. Note that *D* is related to the ΔG of the extraction process.

Sometimes, the distribution ratio is referred to as the partition coefficient, which is often expressed as the logarithm. Note that a distribution ratio for uranium and neptunium between two inorganic solids (zirconolite and perovskite) has been reported. In solvent extraction, two immiscible liquids are shaken together. The more polar solutes dissolve preferentially in the more polar solvent, and the less polar solutes in the less polar solvent. In this experiment, the nonpolar halogens preferentially dissolve in the nonpolar mineral oil.

After performing liquid–liquid extraction, a quantitative measure must be taken to determine the ratio of the solution's total concentration in each phase of the extraction. This quantitative measure is known as the distribution ratio or distribution coefficient.

Separation Factors

The separation factor is one distribution ratio divided by another; it is a measure of the ability of the system to separate two solutes. For instance, if the distribution ratio for nickel (D_{Ni}) is 10 and the distribution ratio for silver (D_{Ag}) is 100, then the silver/nickel separation factor ($SF_{Ag/Ni}$) is equal to $D_{Ag}/D_{Ni} = SF_{Ag/Ni} = 10$.

Decontamination Factor

This is used to express the ability of a process to remove a contaminant from a product. For instance, if a process is fed with a mixture of 1:9 cadmium to indium, and the product is a 1:99 mixture of cadmium and indium, then the decontamination factor (for the removal of cadmium) of the process is 0.11 / 0.01 = 11.

Slopes of Graphs

The easy way to work out the extraction mechanism is to draw graphs and measure the slopes. If for an extraction system the *D* value is proportional to the square of the concentration of a reagent (*Z*) then the slope of the graph of $\log_{10}(D)$ against $\log_{10}([[Z]])$ will be two.

Measures of Success

Success of liquid–liquid extraction is measured through separation factors and decontamination factors. The best way to understand the success of an extraction column is through the liquid–liquid equilibrium (LLE) data set. The data set can then be converted into a curve to determine the steady state partitioning behavior of the solute between the two phases. The y-axis is the concentration of solute in the extract (solvent) phase, and the x-axis is the concentration of the solute in the raffinate phase. From here, one can determine steps for optimization of the process.

Techniques

Batchwise Single Stage Extractions

This is commonly used on the small scale in chemical labs. It is normal to use a separating funnel. Processes include DLLME and direct organic extraction.

Dispersive Liquid–Liquid Microextraction (DLLME)

A process used to extract small amounts of organic compounds from water samples. This process is done by injecting small amounts of an appropriate extraction solvent (C_2Cl_4) and a disperser solvent (acetone) into the aqueous solution. The resulting solution is then centrifuged to separate the organic and aqueous layers. This process is useful in extraction organic compounds such as organochloride and organophsophorus pesticides, as well as substituted benzene compounds from water samples.

Direct Organic Extraction

By mixing partially organic soluble samples in organic solvent (toluene, benzene, xylene), the organic soluble compounds will dissolve into the solvent and can be separated using a separatory funnel. This process is valuable in the extraction of proteins and specifically phosphoprotein and phosphopeptide phosphatases.

4Another example of this application is extracting anisole from a mixture of water and 5% acetic acid using ether, then the anisole will enter the organic phase. The two phases would then be separated. The acetic acid can then be scrubbed (removed) from the organic phase by shaking the organic extract with sodium bicarbonate. The acetic acid reacts with the sodium bicarbonate to form sodium acetate, carbon dioxide, and water.

Caffeine can also be extracted from coffee beans and tea leaves using a direct organic extraction. The beans or leaves can be soaked in ethyl acetate which favorably dissolves the caffeine, leaving a majority of the coffee or tea flavor remaining in the initial sample.

Multistage Countercurrent Continuous Processes

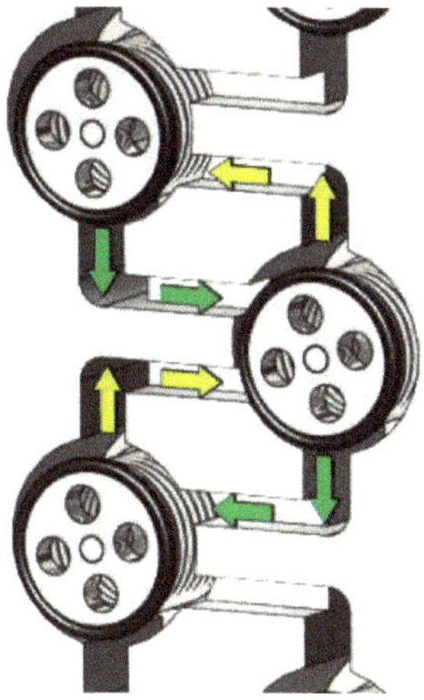

Coflore continuous countercurrent extractor.

These are commonly used in industry for the processing of metals such as the lanthanides; because the separation factors between the lanthanides are so small many extraction stages are needed. In the multistage processes, the aqueous raffinate from one extraction unit is fed to the next unit as the aqueous feed, while the organic phase is moved in the opposite direction. Hence, in this way, even if the separation between two metals in each stage is small, the overall system can have a higher decontamination factor.

Multistage countercurrent arrays have been used for the separation of lanthanides. For the design of a good process, the distribution ratio should be not too high (>100) or too low (<0.1) in the extraction portion of the process. It is often the case that the process will have a section for scrubbing unwanted metals from the organic phase, and finally a stripping section to obtain the metal back from the organic phase.

Mixer–Settlers

Battery of mixer-settlers counter currently interconnected. Each mixer-settler unit provides a single stage of extraction. A mixer settler consists of a first stage that mixes the phases together followed by a quiescent settling stage that allows the phases to separate by gravity.

In the multistage countercurrent process, multiple mixer settlers are installed with mixing and settling chambers located at alternating ends for each stage (since the outlet of the settling sections feed the inlets of the adjacent stage's mixing sections). Mixer-settlers are used when a process requires longer residence times and when the solutions are easily separated by gravity. They require a large facility footprint, but do not require much headspace, and need limited remote maintenance capability for occasional replacement of mixing motors. (Colven, 1956; Davidson, 1957)

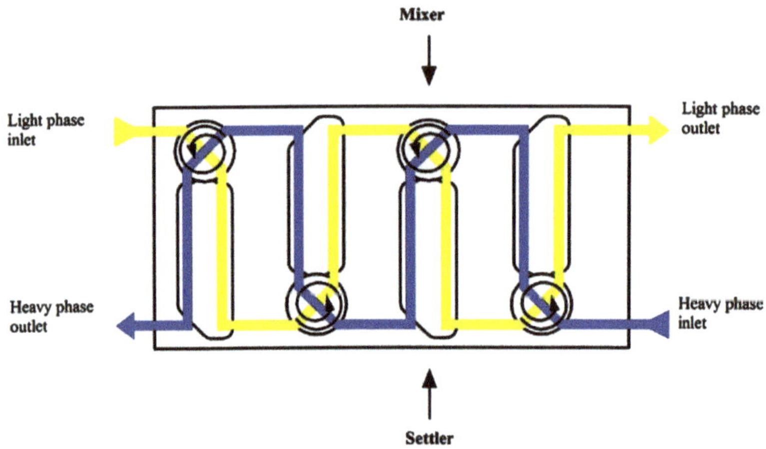

4 stage battery of mixer-settlers for counter-current extraction.

Centrifugal Extractors

Centrifugal extractors mix and separate in one unit. Two liquids will be intensively mixed between the spinning rotor and the stationary housing at speeds up to 6000 RPM. This develops great surfaces for an ideal mass transfer from the aqueous phase into the organic phase. At 200–2000 g, both phases will be separated again. Centrifugal extractors minimize the solvent in the process, optimize the product load in the solvent and extract the aqueous phase completely. Counter current and cross current extractions are easily established.

Extraction without Chemical Change

Some solutes such as noble gases can be extracted from one phase to another without the need for a chemical reaction (see absorption). This is the simplest type of solvent extraction. When a solvent is extracted, two immiscible liquids are shaken together. The more polar solutes dissolve preferentially in the more polar solvent, and the less polar solutes in the less polar solvent. Some solutes that do not at first sight appear to undergo a reaction during the extraction process do not have distribution ratio that is independent of concentration. A classic example is the extraction of carboxylic acids (**HA**) into nonpolar media such as benzene. Here, it is often the case that the carboxylic acid will form a dimer in the organic layer so the distribution ratio will change as a function of the acid concentration (measured in either phase).

For this case, the extraction constant k is described by $k = [[\mathbf{HA}_{\text{organic}}]]^2/[[\mathbf{HA}_{\text{aqueous}}]]$

Solvation Mechanism

Using solvent extraction it is possible to extract uranium, plutonium, or thorium from acid solutions. One solvent used for this purpose is the organophosphate tributyl phosphate (TBP).

The PUREX process that is commonly used in nuclear reprocessing uses a mixture of tri-n-butyl phosphate and an inert hydrocarbon (kerosene), the uranium(VI) are extracted from strong nitric acid and are back-extracted (stripped) using weak nitric acid. An organic soluble uranium complex $[UO_2(TBP)_2(NO_3)_2]$ is formed, then the organic layer bearing the uranium is brought into contact with a dilute nitric acid solution; the equilibrium is shifted away from the organic soluble uranium complex and towards the free TBP and uranyl nitrate in dilute nitric acid. The plutonium(IV) forms a similar complex to the uranium(VI), but it is possible to strip the plutonium in more than one way; a reducing agent that converts the plutonium to the trivalent oxidation state can be added. This oxidation state does not form a stable complex with TBP and nitrate unless the nitrate concentration is very high (circa 10 mol/L nitrate is required in the aqueous phase). Another method is to simply use dilute nitric acid as a stripping agent for the plutonium. This PUREX chemistry is a classic example of a solvation extraction.

Here in this case $D_U = k\ TBP^2[[NO_3]]^2$

Ion Exchange Mechanism

Another extraction mechanism is known as the ion exchange mechanism. Here, when an ion is transferred from the aqueous phase to the organic phase, another ion is transferred in the other direction to maintain the charge balance. This additional ion is often a hydrogen ion; for ion exchange mechanisms, the distribution ratio is often a function of pH. An example of an ion exchange extraction would be the extraction of americium by a combination of terpyridine and a carboxylic acid in *tert*-butyl benzene. In this case

$D_{Am} = k$ terpyridine^{1}carboxylic acid^{3}H+$^{-3}$

Another example is the extraction of zinc, cadmium, or lead by a dialkyl phosphinic acid (R_2PO_2H) into a nonpolar diluent such as an alkane. A non-polar diluent favours the formation of uncharged non-polar metal complexes.

Some extraction systems are able to extract metals by both the solvation and ion exchange mechanisms; an example of such a system is the americium (and lanthanide) extraction from nitric acid by a combination of 6,6'-*bis*-(5,6-dipentyl-1,2,4-triazin-3-yl)-2,2'-bipyridine and 2-bromohexanoic acid in *tert*-butyl benzene. At both high- and low-nitric acid concentrations, the metal distribution ratio is higher than it is for an intermediate nitric acid concentration.

Ion Pair Extraction

It is possible by careful choice of counterion to extract a metal. For instance, if the nitrate concentration is high, it is possible to extract americium as an anionic nitrate complex if the mixture contains a lipophilic quaternary ammonium salt.

An example that is more likely to be encountered by the *'average'* chemist is the use of a phase transfer catalyst. This is a charged species that transfers another ion to the organic phase. The ion reacts and then forms another ion, which is then transferred back to the aqueous phase.

For instance, the 31.1 kJ mol^{-1} is required to transfer an acetate anion into nitrobenzene, while the energy required to transfer a chloride anion from an aqueous phase to nitrobenzene is 43.8 kJ mol^{-1}. Hence, if the aqueous phase in a reaction is a solution of sodium acetate while the organic phase is a nitrobenzene solution of benzyl chloride, then, when a phase transfer catalyst, the acetate anions can be transferred from the aqueous layer where they react with the benzyl chloride to form benzyl acetate and a chloride anion. The chloride anion is then transferred to the aqueous phase. The transfer energies of the anions contribute to that given out by the reaction.

A 43.8 to 31.1 kJ mol^{-1} = 12.7 kJ mol^{-1} of additional energy is given out by the reaction when compared with energy if the reaction had been done in nitrobenzene using one equivalent weight of a tetraalkylammonium acetate.

Types of Aqueous two-Phase Extractions

Polymer–polymer systems. In a Polymer–polymer system, both phases are generated by a dissolved polymer. The heavy phase will generally be Polyethylene glycol (PEG), and the light phase is generally a polysaccharide. Traditionally, the polymer used is dextran. However, dextran is relatively expensive, and research has been exploring using less expensive polysaccharides to generate the light phase. If the target compound being separated is a protein or enzyme, it is possible to incorporate a ligand to the target into one of the polymer phases. This improves the target's affinity to that phase, and improves its ability to partition from one phase into the other. This, as well as the absence of solvents or other denaturing agents, makes polymer–polymer extractions an attractive option for purifying proteins. The two phases of a polymer–polymer system often have very similar densities, and very low surface tension between them. Because of this, demixing a polymer–polymer system is often much more difficult than demixing a solvent extraction. Methods to improve the demixing include centrifugation, and application of an electric field.

Polymer–salt systems. Aqueous two-phase systems can also be generated by introducing a high concentration of salt to a polymer solution. The polymer phase used is generally still PEG. Generally, a kosmotropic salt, such as Na_3PO_4 is used, however PEG–NaCl systems have been documented when the salt concentration is high enough. Since polymer–salt systems demix readily they are easier to use. However, at high salt concentrations, proteins generally either denature, or precipitate from solution. Thus, polymer–salt systems are not as useful for purifying proteins.

Ionic liquids systems. Ionic liquids are ionic compounds with low melting points. While they are not technically aqueous, recent research has experimented with using them in an extraction that does not use organic solvents.

Applications

- DNA purification: The ability to purify DNA from a sample is important for many modern biotechnology processes. However, samples often contain nucleases that degrade the target DNA before it can be purified. It has been shown that

DNA fragments will partition into the light phase of a polymer–salt separation system. If ligands known to bind and deactivate nucleases are incorporated into the polymer phase, the nucleases will then partition into the heavy phase and be deactivated. Thus, this polymer–salt system is a useful tool for purifying DNA from a sample while simultaneously protecting it from nucleases.

- Food industry: The PEG–NaCl system has been shown to be effective at partitioning small molecules, such as peptides and nucleic acids. These compounds are often flavorants or odorants. The system could then be used by the food industry to isolate or eliminate particular flavors. Caffeine extraction used to be done using liquid–liquid extraction, specifically direct and indirect liquid–liquid extraction (Swiss Water Method), but has since moved towards super-critical CO_2 as it is cheaper and can be done on a commercial scale.
- Analytical chemistry: Often there are chemical species present or necessary at one stage of sample processing that will interfere with the analysis. For example, some air monitoring is performed by drawing air through a small glass tube filled with sorbent particles that have been coated with a chemical to stabilize or derivatize the analyte of interest. The coating may be of such a concentration or characteristics that it would damage the instrumentation or interfere with the analysis. If the sample can be extracted from the sorbent using a nonpolar solvent (such as toluene or carbon disulfide), and the coating is polar (such as HBr or phosphoric acid) the dissolved coating will partition into the aqueous phase. Clearly the reverse is true as well, using polar extraction solvent and a nonpolar solvent to partition a nonpolar interferent. A small aliquat of the organic phase (or in the latter case, polar phase) can then be injected into the instrument for analysis.
- Purification of amines: Amines (analogously to ammonia) have a lone pair of electrons on the nitrogen atom that can form a relatively weak bond to a hydrogen atom. It is therefore the case that under acidic conditions amines are typically protonated, carrying a positive charge and under basic conditions they are typically deprotonated and neutral. Amines of sufficiently low molecular weight are rather polar and can form hydrogen bonds with water and therefore will readily dissolve in aqueous solutions. Deprotonated amines on the other hand, are neutral and have *greasy*, nonpolar organic substituents, and therefore have a higher affinity for nonpolar inorganic solvents. As such purification steps can be carried out where an aqueous solution of an amine is neutralized with a base such as sodium hydroxide, then shaken in a separatory funnel with a nonpolar solvent that is immiscible with water. The organic phase is then drained off. Subsequent processing can recover the amine by techniques such as recrystallization, evaporation or distillation; subsequent extraction back to a polar phase can be performed by adding HCl and shaking again in a separatory funnel (at which point the ammonium ion could be recovered by adding an insoluble counterion), or in either phase, reactions could be performed as part of a chemical synthesis.

Kinetics of Extraction

It is important to investigate the rate at which the solute is transferred between the two phases, in some cases by an alteration of the contact time it is possible to alter the selectivity of the extraction. For instance, the extraction of palladium or nickel can be very slow because the rate of ligand exchange at these metal centers is much lower than the rates for iron or silver complexes.

Aqueous Complexing Agents

If a complexing agent is present in the aqueous phase then it can lower the distribution ratio. For instance, in the case of iodine being distributed between water and an inert organic solvent such as carbon tetrachloride then the presence of iodide in the aqueous phase can alter the extraction chemistry.

Instead of $D_{I^{+2}}$ being a constant it becomes $D_{I^{+2}}$

This is because the iodine reacts with the iodide to form I_3^-. The I_3^- anion is an example of a polyhalide anion that is quite common.

Industrial Process Design

In a typical scenario, an industrial process will use an extraction step in which solutes are transferred from the aqueous phase to the organic phase; this is often followed by a scrubbing stage in which unwanted solutes are removed from the organic phase, then a stripping stage in which the wanted solutes are removed from the organic phase. The organic phase may then be treated to make it ready for use again.

After use, the organic phase may be subjected to a cleaning step to remove any degradation products; for instance, in PUREX plants, the used organic phase is washed with sodium carbonate solution to remove any dibutyl hydrogen phosphate or butyl dihydrogen phosphate that might be present.

Equipment

While solvent extraction is often done on a small scale by synthetic lab chemists using a separatory funnel, Craig apparatus or membrane-based techniques, it is normally done on the industrial scale using machines that bring the two liquid phases into contact with each other. Such machines include centrifugal contactors, Thin Layer Extraction, spray columns, pulsed columns, and mixer-settlers.

A typical modular process system for liquid–liquid extraction includes processing equipment such as columns, reactors, heat exchangers and pumps, mounted within a strong structural steel frame. Additional features of large-scale modular process systems include thermal insulation, lighting, safety showers, control systems, and fire protection systems.

Phase separation during a laboratory scale LLE. The upper organic ether solution of MTBE is being extracted with the lower alkaline aqueous sodium bicarbonate solution to remove benzoic acid as the benzoate anion, leaving a non-acidic organic, benzil, (yellow in color) in the organic phase.

SOLID PHASE EXTRACTION

Solid-phase extraction (SPE) is a sample preparation process by which compounds that are dissolved or suspended in a liquid mixture are separated from other compounds in the mixture according to their physical and chemical properties. Analytical laboratories use solid phase extraction to concentrate and purify samples for analysis. Solid phase extraction can be used to isolate analytes of interest from a wide variety of matrices, including urine, blood, water, beverages, soil, and animal tissue.

SPE uses the affinity of solutes dissolved or suspended in a liquid (known as the mobile phase) for a solid through which the sample is passed (known as the stationary phase) to separate a mixture into desired and undesired components. The result is that either the desired analytes of interest or undesired impurities in the sample are retained on the stationary phase. The portion that passes through the stationary phase is collected or discarded, depending on whether it contains the desired analytes or undesired impurities. If the portion retained on the stationary phase includes the desired analytes, they can then be removed from the stationary phase for collection in an additional step, in which the stationary phase is rinsed with an appropriate eluent.

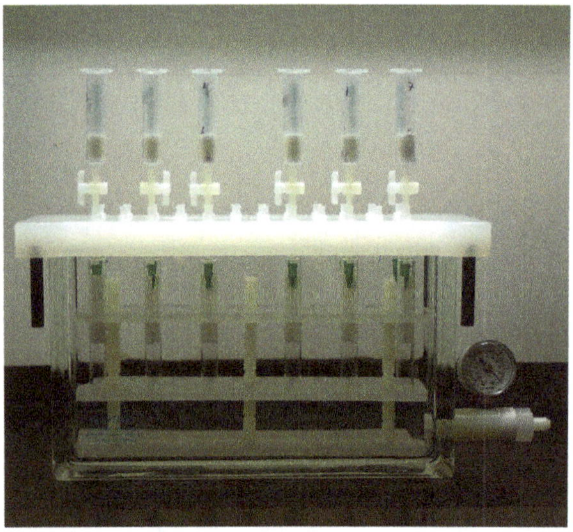

A typical solid phase extraction manifold. The cartridges drip into the chamber below, where tubes collect the effluent. A vacuum port with gauge is used to control the vacuum applied to the chamber.

Normal Phase SPE Procedure

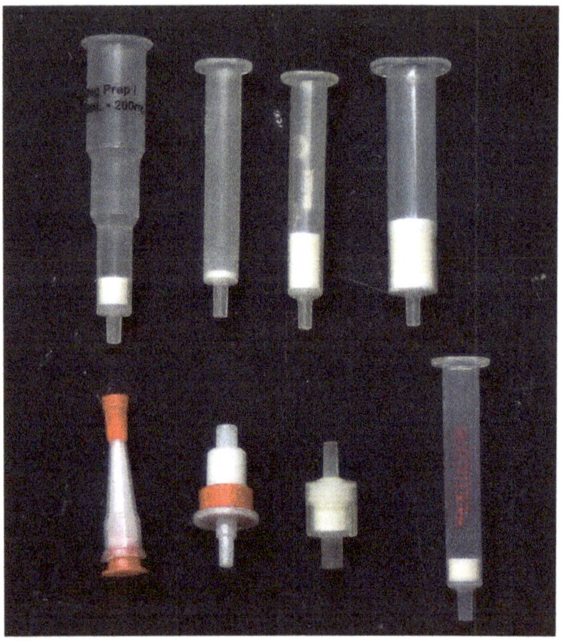

A selection of solid phase extraction cartridges, available in many sizes, shapes, and types of stationary phase.

A typical solid phase extraction involves five basic steps. First, the cartridge is equilibrated with a non-polar or slightly polar solvent, which wets the surface and penetrates the bonded phase. Then water, or buffer of the same composition as the sample, is typically washed through the column to wet the silica surface. The sample is then added to the cartridge. As the sample passes through the stationary phase, the polar analytes in the sample will interact and retain on the polar sorbent while the solvent, and other non-polar impurities pass through the cartridge. After the sample is loaded, the cartridge is washed with a non-polar solvent to remove further impurities. Then, the analyte is eluted with a polar solvent or a buffer of the appropriate pH.

A stationary phase of polar functionally bonded silicas with short carbons chains frequently makes up the solid phase. This stationary phase will adsorb polar molecules which can be collected with a more polar solvent.

Reversed Phase SPE

Reversed phase SPE separates analytes based on their polarity. The stationary phase of a reversed phase SPE cartridge is derivatized with hydrocarbon chains, which retain compounds of mid to low polarity due to the hydrophobic effect. The analyte can be eluted by washing the cartridge with a non-polar solvent, which disrupts the interaction of the analyte and the stationary phase.

A stationary phase of silicon with carbon chains is commonly used. Relying on mainly non-polar, hydrophobic interactions, only non-polar or very weakly polar compounds will adsorb to the surface.

Ion Exchange SPE

Ion exchange sorbents separate analytes based on electrostatic interactions between the analyte of interest and the positively or negatively charged groups on the stationary phase. For ion exchange to occur, both the stationary phase and sample must be at a pH where both are charged.

Anion Exchange

Anion exchange sorbents are derivatized with positively charged functional groups that interact and retain negatively charged anions, such as acids. Strong anion exchange sorbents contain quaternary ammonium groups that have a permanent positive charge in aqueous solutions, and weak anion exchange sorbents use amine groups which are charged when the pH is below about 9. Strong anion exchange sorbents are useful because any strongly acidic impurities in the sample will bind to the sorbent and usually will not be eluted with the analyte of interest; to recover a strong acid a weak anion exchange cartridge should be used. To elute the analyte from either the strong or weak sorbent, the stationary phase is washed with a solvent that neutralizes the charge of either the analyte, the stationary phase, or both. Once the charge is neutralized, the electrostatic interaction between the analyte and the stationary phase no longer exists and the analyte will elute from the cartridge.

Cation Exchange

Cation exchange sorbents are derivatized with functional groups that interact and retain positively charged anions, such as bases. Strong cation exchange sorbents contain aliphatic sulfonic acid groups that are always negatively charged in aqueous solution, and weak cation exchange sorbents contain aliphatic carboxylic acids, which are charged when the pH is above about 5. Strong cation exchange sorbents are useful because any strongly basic impurities in the sample will bind to the sorbent and usually will not be eluted with the analyte of interest; to recover a strong base a weak cation exchange cartridge should be used. To elute the analyte from either the strong or weak sorbent, the stationary phase is washed with a solvent that neutralizes ionic interaction between the analyte and the stationary phase.

Cartridges

The stationary phase comes in the form of a packed syringe-shaped cartridge, a 96 well plate, a 47- or 90-mm flat disk, or a microextraction by packed sorbent (MEPS) device, a SPE method that uses a packed sorbent material in a liquid handling syringe. These can be mounted on its specific type of extraction manifold. The manifold allows multiple samples to be processed by holding several SPE media in place and allowing for an equal number of samples to pass through them simultaneously. A typical cartridge SPE manifold can accommodate up to 24 cartridges, while a typical disk SPE manifold can accommodate 6 disks. Most SPE manifolds are equipped with a vacuum port. Application of vacuum speeds up the extraction process by pulling the liquid sample through the stationary phase. The analytes are collected in sample tubes inside or below the manifold after they pass through the stationary phase.

Solid phase extraction cartridges and disks are available with a variety of stationary phases, each of which can separate analytes according to different chemical properties. Most stationary phases are based on silica that has been bonded to a specific functional group. Some of these functional groups include hydrocarbon chains of variable length (for reversed phase SPE), quaternary ammonium or amino groups (for anion exchange), and sulfonic acid or carboxyl groups (for cation exchange).

Solid-Phase Microextraction

Solid-phase microextraction (SPME), is a solid phase extraction technique that involves the use of a fiber coated with an extracting phase, that can be a liquid (polymer) or a solid (sorbent), which extracts different kinds of analytes (including both volatile and non-volatile) from different kinds of media, that can be in liquid or gas phase. The quantity of analyte extracted by the fibre is proportional to its concentration in the sample as long as equilibrium is reached or, in case of short time pre-equilibrium, with help of convection or agitation.

FIELD FLOW FRACTIONATION

Field-flow fractionation, abbreviated FFF, is a separation technique where a field is applied to a fluid suspension or solution pumped through a long and narrow channel, perpendicular to the direction of flow, to cause separation of the particles present in the fluid, depending on their differing "mobilities" under the force exerted by the field. It was invented and first reported by J. Calvin Giddings. The method of FFF is unique to other separation techniques due to the fact that it can separate materials over a wide colloidal size range while maintaining high resolution. Although FFF is an extremely versatile technique, there is no "one size fits all" method for all applications.

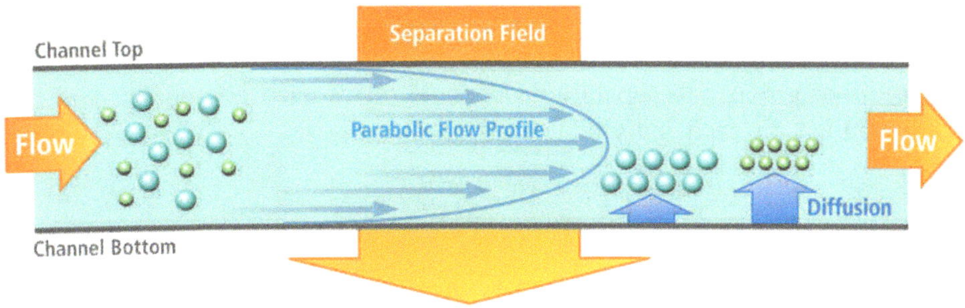

Flow field-flow fractionation (AF4) channel cross section, where the rate of laminar flow within the channel is not uniform. It travels in a parabolic pattern with the speed of the flow, increasing towards the centre of the channel and decreasing towards the sides.

In field-flow fractionation the *field* can be asymmetrical *flow* through a semi-permeable membrane, gravitational, centrifugal, thermal-gradient, electrical, magnetic etc. In all cases, the separation mechanism is born from differences in particle mobility (electrophoretic, when the field is a DC electric field causing a transverse electric current flow) under the forces of the field, *in equilibrium with* the forces of diffusion: an often-parabolic laminar-flow-velocity profile in the channel determines the velocity of a particular particle, based on its equilibrium position from the wall of the channel. The ratio of the velocity of a species of particle to the average velocity of the fluid is called the *retention ratio*.

Fundamental Principles

Field flow fractionation is based on laminar flow of particles in a solution. These sample components will change levels and speed based on their size/mass. Since these components will be travelling at different speeds, separation occurs. A simplified explanation of the setup is as follows. The sample separation occurs in a thin, ribbon-like, channel in which there is an inlet flow and a perpendicular field flow. The inlet flow is where the carrier liquid is pumped into the channel and it creates a parabolic flow profile and it propels the sample towards the outlet of the channel.

Relating Force (F) to Retention Time (TR)

The relationship between the separative force field and retention time can be illustrated from first principles. Consider two particle populations within the FFF channel. The cross field drives both particle clouds towards the bottom "accumulation" wall. Opposing this force field is the particles natural diffusion, or Brownian motion, which produces a counter acting motion. When these two transport process reach equilibrium the particle concentration c approaches the exponential function of elevation x above the accumulation wall as illustrated in equation 1.

$$c = c_0 e^{\frac{-x}{l}}$$

l represents the characteristic elevation of the particle cloud. This relates to the height that the particle group can reach within the channel and only when the value for l is different for either group will separation occur. The l of each component can be related to the force applied on each individual particle.

$$l = \frac{kT}{F}$$

Where k is the Boltzmann constant, T is absolute pressure and F is the force exerted on a single particle by the cross flow. This shows how the characteristic elevation value is inversely dependent to the Force applied. Therefore, F governs the separation process. Hence, by varying the field strength the separation can be controlled to achieve optimal levels. The velocity V of a cloud of molecules is simply the average velocity of an exponential distribution embedded in a parabolic flow profile. Retention time, tr can be written as:

$$t_r = \frac{L}{V}$$

Where L is the channel length. Subsequently, the retention time can be written as:

$t_r/t^o = w/6l\ \lfloor \coth w/2l - 2l/w \rfloor^{-1}$

Where to is the void time (emergence of a non-retained tracer) and w is the sample thickness. Substituting in kT/F in place of l illustrates the retention time with respect to the cross force applied.

$t_r/t^o = Fw/6kT\ \lfloor \coth Fw/2kT - 2kT/Fw \rfloor^{-1}$

For an efficient operation the channel thickness value w far exceeds l. When this is the case the term in the brackets approaches unity. Therefore, equation 5 can be approximated as:

$t_r/t^o = w/6l = Fw/6kT$

Thus tr is roughly proportional to F. The separation of particle bands X and Y, represented by the finite increment Δtr in their retention times, is achieved only if the force increment ΔF between them is sufficient. A differential in force of only 10–16 N is required for this to be the case. The magnitude of F and ΔF depend on particle properties, field strength and the type of field. This allows for variations and specialisations for the technique. From this basic principle many forms of FFF have evolved varying by the nature of the separative force applied and the range in molecule size to which they are targeted.

Fractogram

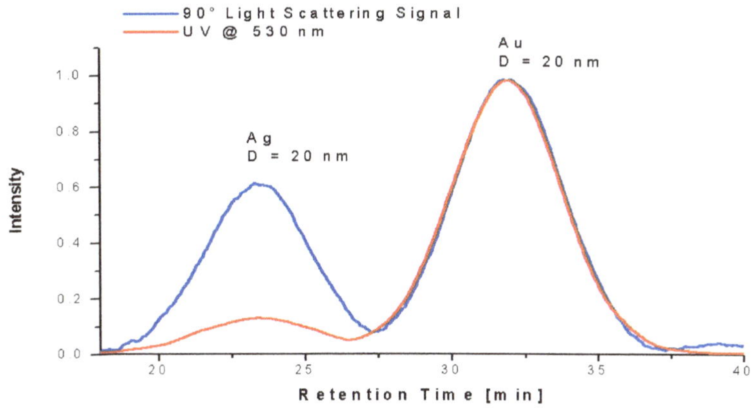

Centrifugal FFF separates by mass (i.e. a combination of particle density and particle size). For example, gold and silver nanoparticles of identical size can be separated into two peaks, according to differences in density of gold and silver.

A graph of detection signal vs. time from an FFF separation, in which various substances present in a fluid are separated based on their flow velocities under some applied external field, such as an flow, centrifugal, thermal or electric field.

Often these substances are particles initially suspended in a small volume of a liquid buffer and pushed along the FFF channel by the buffer. The varying velocities of a particular species of particles may be due to its size, its mass, and/or its distance from the walls of a channel with non-uniform flow-velocity. The presence of different species in a sample can thus be identified through detection of a common property at some distance down the long channel, and by the resulting fractogram indicating the presence of the various species by peaks, due to the different times of arrival characteristic of each species and its physical and chemical properties.

In an electrical FFF, an electric field controls the velocity by controlling the lateral position of either a charged (having electrophoretic mobility) or polarized (being levitated in a non-uniform field) species in a capillary channel with a hydrodynamically parabolic flow-velocity profile, meaning that the velocity of the pumped fluid is highest midway between the walls of the channel and it monotonically decays to a minimum of zero at the wall surface.

Forms of FFF

Most techniques available today are advances on those originally created by Prof. Giddings nearly 4 decades ago.

Flow FFF

Of these techniques Flow FFF was the first to be offered commercially. Flow FFF separates particles based on size, independent of density and can measure macromolecules in the range of 1 nm to 1 μm. In this respect it is the most versatile FFF sub-technique available. The cross flow in Flow FFF enters through a porous frit at the top of the channel, exiting through a semi-permeable membrane outlet frit on the accumulation wall (i.e. the bottom wall).

Hollow Fiber Flow FFF

Hollow fiber flow FFF (HF5) has been pioneered by H.L. Lee, J.F.G. Reis and E. N. Lightfoot. HF5 has been applied towards the analysis of lattices and other macromolecules. HF5 was the first form of flow FFF (1974) to be developed. Flat membranes soon outperformed hollow fibers and forced HF5 into obscurity. One of the drawbacks of HF5 is the availability of the membranes with uniform pore sizes. There are different kinds of ceramic and polymeric hollow fiber membranes used in practice.

Asymmetric Flow FFF (AF4)

Asymmetric Flow FFF (AF4), on the other hand, has only one semi-permeable membrane on the bottom wall of the channel. The cross flow is, therefore, created by the carrier liquid exiting the bottom of the channel. This offers an extremely gentle separation and an "ultra-broad" separation range. High Temperature Asymmetric Flow Field-Flow Fractionation is the most advanced technology for the separation of high and ultra-high molar mass polymers, macromolecules and nanoparticles in the size range.

Thermal FFF

Thermal FFF, as the name suggests, establishes a separation force by applying a temperature gradient to the channel. The top channel wall is heated and the bottom wall is cooled driving polymers and particles towards the cold wall by thermal diffusion. Thermal FFF was developed as a technique for separating synthetic polymers in organic solvents. Thermal FFF is unique amongst FFF techniques in that it can separate macromolecules by both molar mass and chemical composition, allowing for the separation of polymer fractions with the same molecular weight. Today this technique is ideally suited for the characterization of polymers, gels and nanoparticles.

Split Flow thin-Cell Fractionation (SPLITT)

Split flow thin-cell fractionation (SPLITT) is a special preparative FFF technique, using gravity for separation of μm-sized particles on a continuous basis. SPLITT is performed by pumping the sample containing liquid into the top inlet at the start of the channel, whilst simultaneously pumping a carrier liquid into the bottom inlet. By controlling the flow rate ratios of the two inlet streams and two outlet streams, the separation can be controlled and the sample separated into two distinct sized fractions. The use of gravity alone as the separating force makes SPLITT the least sensitive FFF technique, limited to particles above 1 μm.

Centrifugal FFF

In centrifugal FFF, the separation field is generated via a centrifugal force. The channel takes the form of a ring, which spins at speeds up to 4900 rpm in the case of Postnova Analytics CF2000 instrument. The flow and sample are pumped into the channel and centrifuged, allowing the operator to resolve the particles by mass (size and density). The advantage of centrifugal FFF lies in the high size resolution that can be achieved by varying the force applied, since particle size is proportional to particle mass to the third power.

The unique advantage presented by centrifugal FFF comes from the techniques capability for high resolution. The only commercially available centrifugal FFF instrument is Postnova Analytics' CF2000, incorporating the unique feature of separating particles by both particle size and density. This allows for the separation of particles with only a 5% difference in size. Centrifugal FFF has the advantage that particles and macromolecules can be separated by particle density, rather than just particle size. In this instance, two identically sized gold and silver nanoparticles can be separated into two peaks, according to differences in density in the gold and silver nanoparticles, separated with the Centrifugal FFF Postnova CF2000 instrument with detection by Dynamic Light Scattering (DLS).

In AF4 separations, the ratio of mass to time is 1:1. With the addition of the third parameter of density to centrifugal FFF, this produces a ratio more akin to mass:time to the power of three. This results in a significantly larger distinction between peaks and result in a greatly improved resolution. This can be particularly useful for novel products, such as composite materials and coated polymers containing nanoparticles, i.e particles which may not vary in size but which do vary in density. In this way two identically sized particles can still be separated into two peaks, providing that the density is different.

REFERENCES

1. Oliva P. Canencia; angelo Mark P. Walag (2016). "Coal Combustion from Power Plant Industry Inmisamis Oriental, Philippines: a Potential Groundwater Contamination and Heavy Metal Detection". Asian Journal of Microbiology, Biotechnology & Environmental Sciences. 18 (1): 55–59. ISSN 0972-3005. Retrieved 2016-04-22.

2. Geankoplis, Christie (2004). Transport Process and Separation Principles. NJ: Pretence Hall. pp. 802–817. ISBN 978-0-13-101367-4.
3. Seader and Henley, J.D. and Ernest J. (2001). Separation Process Principles. U.K: John Wiley & Sons inc. pp. 639–641. ISBN 978-0-471-46480-8.
4. Li, Li; Jing Ge; Renjie Chen; Feng Wu; Shi Chen; Xiaoxiao Zhang (March 16, 2010). "Environmental friendly leaching reagent for cobalt and lithium recovery". International Journal of Integrated Waste Management, Science and Technology. Waste Management. 30 (12): 2615–2621. doi:10.1016/j.wasman.2010.08.008. Retrieved Nov 2011. Check date values in: |access-date= (help)
5. Li, Li; Jing Ge; Feng Wu; Renjie Chen; Shi Chen; Borong Wu (2010-04-15). "Recovery of cobalt and lithium from spent lithium ion batteries using organic citric acid as leachant". Journal of Hazardous Materials. 176 (1-3): 288–293. doi:10.1016/j.jhazmat.2009.11.026. Retrieved Nov 2011.
6. (PDF) https://web.archive.org/web/20080309151803/http://www-ssrl.slac.stanford.edu/pubs/activity_rep/ar98/2370-vance.pdf. Archived from the original (PDF) on March 9, 2008. Retrieved January 21, 2006. Missing or empty |title= (help)
7. pnjjrose. "Solvent Extraction Notes".
8. http://courses.chem.psu.edu/chem36/Experiments/PDF's_for_techniques/Liquid_Liquid.pdf
9. "Basic Technology and Tools in Chemical Engineering Field - S. Wesley - Documents".
10. http://modularprocess.com/wp-content/uploads/2015/07/Article-Liquid-LiquidExtraction-ChE-2004.pdf
11. Rezaee, Mohammad; Assadi, Yaghoub; Milani Hosseini, Mohammad-Reza; Aghaee, Elham; Ahmadi, Fardin; Berijani, Sana (2006). "Determination of organic compounds in water using dispersive liquid–liquid microextraction". Journal of Chromatography A. 1116 (1-2): 1–9. doi:10.1016/j.chroma.2006.03.007. ISSN 0021-9673.
12. Shacter, Emily (1984). "Organic extraction of Pi with isobutanol/toluene". Analytical Biochemistry. 138 (2): 416–420. doi:10.1016/0003-2697(84)90831-5. ISSN 0003-2697.
13. Senese F (20 September 2005). "How is coffee decaffeinated?". General Chemistry Online. Retrieved 3 August 2009.
14. Binnemans, Koen (2007). "Lanthanides and Actinides in Ionic Liquids". Chemical Reviews. 107 (6): 2592–2614. doi:10.1021/cr050979c. ISSN 0009-2665.
15. Liquid–Liquid Extraction Equipment, Jack D. Law and Terry A. Todd, Idaho National Laboratory.
16. "Riegel's Handbook of Industrial Chemistry".
17. Scholz, F.; S. Komorsky-Lovric; M. Lovric (February 2000). "A new access to Gibbs energies of transfer of ions across liquid|liquid interfaces and a new method to study electrochemical processes at well-defined three-phase junctions".

Electrochemistry Communications. Elsevier. 2 (2): 112–118. doi:10.1016/S1388-2481(99)00156-3.

18. Danil de Namor, A.F.; T. Hill (1983). Journal of the Chemical Society, Faraday Transactions: 2713. Missing or empty |title= (help)
19. zamani, Dariush. “Extraction Operation”.
20. Peker, Hulya; Srinivasan, M. P.; Smith, J. M.; McCoy, Ben J. (1992). “Caffeine extraction rates from coffee beans with supercritical carbon dioxide”. AIChE Journal. 38 (5): 761–770. doi:10.1002/aic.690380513. ISSN 0001-1541.
21. Emden, Lorenzo. “Decaffeination 101: Four Ways to Decaffeinate Coffee”. Coffee Confidential. Retrieved 29 October 2014.
22. Adamo, Andrea; Heider, Patrick L.; Weeranoppanant, Nopphon; Jensen, Klavs F. (2013). “Membrane-Based, Liquid–Liquid Separator with Integrated Pressure Control”. Industrial & Engineering Chemistry Research. 52 (31): 10802–10808. doi:10.1021/ie401180t. ISSN 0888-5885.
23. http://modularprocess.com/wp-content/uploads/2015/06/Article-MovingToModular-ChE-01-2015.pdf
24. Mackenzie, Murdoch. “The Solvent Extraction of Some Major Metals” (PDF). Cognis GmbH. Retrieved 2008-11-18.
25. Filiz, M.; Sayar, N.A.; Sayar, A.A. (2006). “Extraction of cobalt(II) from aqueous hydrochloric acid solutions into alamine 336–m-xylene mixtures”. Hydrometallurgy. 81 (3-4): 167–173. doi:10.1016/j.hydromet.2005.12.007. ISSN 0304-386X.
26. Baba, Yoshinari; Iwakuma, Minako; Nagami, Hideto (2002). “Extraction Mechanism for Copper(II) with 2-Hydroxy-4-n-octyloxybenzophenone Oxime”. Industrial & Engineering Chemistry Research. 41 (23): 5835–5841. doi:10.1021/ie0106736. ISSN 0888-5885.
27. Sanchez, J.M.; Hidalgo, M.; Salvadó, V.; Valiente, M. (1999). “EXTRACTION OF NEODYMIUM(III) AT TRACE LEVEL WITH DI(2-ETHYL-HEXYL) PHOSPHORIC ACID IN HEXANE”. Solvent Extraction and Ion Exchange. 17 (3): 455–474. doi:10.1080/07366299908934623. ISSN 0736-6299.
28. Lee W. John. “A Potential Nickel / Cobalt Recovery Process”. BioMetallurgical Pty Ltd.
29. “Precious Metals Refining By Solvent Extraction”. Halwachs Edelmetallchemie und Verfahrenstechnik. Retrieved 2008-11-18.
30. Giridhar, P.; Venkatesan, K.A.; Srinivasan, T.G.; Vasudeva Rao, P.R. (2006). “Extraction of fission palladium by Aliquat 336 and electrochemical studies on direct recovery from ionic liquid phase”. Hydrometallurgy. 81 (1): 30–39. doi:10.1016/j.hydromet.2005.10.001. ISSN 0304-386X.
31. Schulz, Wallace W.; Schiefelbein, Gary F.; Bruns, Lester E. (1969). “Pyrochemical Extraction of Polonium from Irradiated Bismuth Metal”. Ind. Eng. Chem. Process Des. Dev. 8 (4): 508–515. doi:10.1021/i260032a013.

32. K. Takeshita; K. Watanabe; Y. Nakano; M. Watanabe (2003). "Solvent extraction separation of Cd(II) and Zn(II) with the organophosphorus extractant D2EHPA and the aqueous nitrogen-donor ligand TPEN". Hydrometallurgy. 70: 63–71. doi:10.1016/s0304-386x(03)00046-x.
33. Hennion, Marie-Claire (1999). "Solid-phase extraction: method development, sorbents, and coupling with liquid chromatography". Journal of Chromatography A. 856 (1-2): 3–54. doi:10.1016/S0021-9673(99)00832-8. ISSN 0021-9673.
34. Augusto, Fabio; Hantao, Leandro W.; Mogollón, Noroska G.S.; Braga, Soraia C.G.N. (2013). "New materials and trends in sorbents for solid-phase extraction". TrAC Trends in Analytical Chemistry. 43: 14–23. doi:10.1016/j.trac.2012.08.012. ISSN 0165-9936.
35. Supelco (1998), Guide to Solid Phase Extraction (PDF)
36. Abdel-Rehim, Mohamed (2011). "Microextraction by packed sorbent (MEPS): A tutorial". Analytica Chimica Acta. 701 (2): 119–128. doi:10.1016/j.aca.2011.05.037. ISSN 0003-2670.
37. M. Abdel-Rehim, AstraZeneca Application "Syringe for solid phase microextraction", Current Patents Gazette, week 0310, WO 03019149, p. 77, (2003).
38. Mitra, Somenath, ed. (2003). Sample Preparation Techniques in Analytical Chemistry. Wiley-Interscience. p. 113.
39. Giddings, JC, FJ Yang, and MN Myers. "Flow Field-Flow Fractionation: a versatile new separation method." Science 193.4259 (1976): 1244–1245.
40. Madou, Marc (2001). Fundamentals of Microfabrication. US: CRC. pp. 565–571. ISBN 0-8493-0826-7.
41. H.L. Lee, J.F.G. Reis and E. N. Lightfoot, AIChE Journal edition 20, 1974, page 776 entitled Single-phase chromatography: Solute retardation by ultrafiltration and electrophoresis

CHAPTER 7

The Technique of Froth Flotation

Chapter Contents

- **Introduction**
- **Dissolved Air Flotation**
- **Deinking**
- **Flocculation**
- **Fractional Distillation**
- **Fractional Freezing**

This method is used for concentration of sulphide ores of copper, lead and zinc. In this method, powdered ore is put in a tank full of water. And then, some Pine oil is added to it. In the tank the particles of sulphide ore are wetted by pine oil whereas the gangue particles are wetted bywater. Then air is passed through this mixture. This results in the agitation of water in tank, which cause the sulphide ore particles to stick with the oil and rise to the surface in the form of froth. The gangue particles being heavier remain behind at the bottom of water tank. The froth is separated and concentrated sulphide ore is obtained from it.

INTRODUCTION

Froth flotation is a process for selectively separating hydrophobic materials from hydrophilic. This is used in mineral processing, paper recycling and waste-water treatment industries. Historically this was first used in the mining industry, where it was one of the great enabling technologies of the 20th century. It has been described as "the single most important operation used for the recovery and upgrading of sulfide ores". The development of froth flotation has improved the recovery of valuable minerals, such as copper- and lead-bearing minerals. Along with mechanized mining, it has allowed the economic recovery of valuable metals from much lower grade ore than previously.

Diagram of a cylindrical froth flotation cell with camera and light used in image analysis of the froth surface.

Initially, naturally occurring chemicals such as fatty acids and oils were used as flotation reagents in a large quantity to increase the hydrophobicity of the valuable minerals. Since then, the process has been adapted and applied to a wide variety of materials to be separated, and additional collector agents, including surfactants and synthetic compounds have been adopted for various applications.

William Haynes patented a process in 1869 for separating sulfide and gangue minerals using oil and called it *bulk-oil flotation*. In 1885 Carrie Everson expanded upon this and patented a process calling for oils but also an acid or a salt.

The first successful commercial flotation process for mineral sulphides was invented by Frank Elmore who worked on the development with his brother, Stanley. The Glasdir copper mine at Llanelltyd, near Dolgellau in North Wales was bought in 1896 by the Elmore brothers in conjunction with their father, William. In 1897, the Elmore brothers installed the world's first industrial size commercial flotation process for mineral beneficiation at the Glasdir mine. The process was not froth flotation but used oil to agglomerate (make balls of) pulverised sulphides and buoy them to the surface, and was patented in 1898 with a description of the process published in 1903 in the *Engineering and Mining Journal*. By this time they had recognized the importance of air bubbles in assisting the oil to carry away the mineral particles. The Elmores had formed a company known as the Ore Concentration Syndicate Ltd to promote the commercial use of the process worldwide. However developments elsewhere, particularly in Australia by Minerals Separation Ltd., led to decades of hard fought legal battles and litigations which, ultimately, were lost as the process was superseded by more advanced techniques. Charles Butters, beginning about 1899, and working with both the Elmores and Minerals Separation's representative E.H. Nutter developed what was known to contemporaries as the "Butters Process". The flotation process was independently invented in the early 1900s in Australia by Charles

Vincent Potter and around the same time by Guillaume Daniel Delprat. This process (developed circa 1902) did not use oil, but relied upon flotation by the generation of gas formed by the introduction of acid into the pulp. In 1902, Froment combined oil and gaseous flotation using a modification of the Potter-Delprat process.

Another process was developed in 1902 by Cattermole, who emulsified the pulp with a small quantity of oil, subjected it to violent agitation, then slow stirring which coagulated the target minerals into nodules which were separated from the pulp by gravity. This was the basis of the Minerals Separation Ltd. process. By 1904, the MacQuisten process (a surface tension based method) was developed but this would not work when slimes were present. In 1912, James M. Hyde modified the Minerals Separation Process and installed it in the Butte and Superior Mill in Basin, Montana, the first such installation in the USA.

John M. Callow, of General Engineering of Salt Lake City, had followed flotation from technical papers and the introduction in both the Butte and Superior Mill, and at Inspiration Copper in Arizona and determined that mechanical agitation was a drawback to the existing technology. Introducing a porous brick with compressed air, and a mechanical stirring mechanism, Callow applied for a patent in 1914. This method, known as Pneumatic Flotation, was recognized to revolutionize the process of flotation concentration. The American Institute of Mining Engineers presented Callow the James Douglas Gold Medal in 1926 for his contributions to the field of flotation.

In the 1960s the froth flotation technique was adapted for deinking recycled paper.

Industries

Mining

Froth flotation to separate plastics, Argonne National Laboratory

Froth flotation cells to concentrate copper and nickel sulfide minerals, Falconbridge, Ontario.

Froth flotation is a process for separating minerals from gangue by taking advantage of differences in their hydrophobicity. Hydrophobicity differences between valuable minerals and waste gangue are increased through the use of surfactants and wetting agents. The selective separation of the minerals makes processing complex (that is, mixed) ores economically feasible. The flotation process is used for the separation of a large range of sulfides, carbonates and oxides prior to further refinement. Phosphates and coal are also upgraded (purified) by flotation technology.

Prior to 1907, nearly all the copper mined in the US came from underground vein deposits, averaging 2.5 percent copper. By 1991, the average grade of copper ore mined in the US had fallen to only 0.6 percent.

Waste Water Treatment

The flotation process is also widely used in industrial waste water treatment plants, where it removes fats, oil, grease and suspended solids from waste water. These units are called dissolved air flotation (DAF) units. In particular, dissolved air flotation units are used in removing oil from the wastewater effluents of oil refineries, petrochemical and chemical plants, natural gas processing plants and similar industrial facilities.

Paper Recycling

Froth flotation is one of the processes used to recover recycled paper. In the paper industry this step is called deinking or just flotation. The target is to release and remove the hydrophobic contaminants from the recycled paper. The contaminants are mostly printing ink and stickies. Normally the setup is a two-stage system with 3,4 or 5 flotation cells in series.

Principle of Operation

Before froth flotation can work, the ore to be treated is reduced to fine particles by crushing and grinding (a process known as comminution) so that the various minerals exist as physically separate grains. This process is known as *liberation*. The particle sizes are typically less than 0.1 mm (100 μm), but sometimes sizes smaller than 7–10 μm are required. There is a tendency for the liberation size of the minerals to decrease over time as the ore bodies with coarse mineral grains that can be separated at larger sizes are depleted and replaced by ore bodies that were formerly considered too difficult.

In the mining industry, the plants where flotation is undertaken to concentrate ore are generally known as *concentrators* or *mills*.

For froth flotation, the ground ore is mixed with water to form a slurry and the desired mineral is rendered hydrophobic by the addition of a surfactant or *collector* chemical (although some mineral surfaces are naturally hydrophobic, requiring little or no addition of collector). The particular chemical depends on the nature of the mineral to be recovered and, perhaps, the natures of those that are not wanted. As an example, sodium ethyl xanthate may be added as a collector in the selective flotation of galena (lead sulfide) to separate it from sphalerite (zinc sulfide). This slurry (more properly called the *pulp*) of hydrophobic particles and hydrophilic particles is then introduced to tanks known as *flotation cells* that are aerated to produce bubbles. The hydrophobic particles attach to the air bubbles, which rise to the surface, forming a froth. The froth is removed from the cell, producing a concentrate (con) of the target mineral.

Frothing agents, known as *frothers*, may be introduced to the pulp to promote the formation of a stable froth on top of the flotation cell.

The minerals that do not float into the froth are referred to as the *flotation tailings* or *flotation tails*. These tailings may also be subjected to further stages of flotation to recover the valuable particles that did not float the first time. This is known as *scavenging*. The final tailings after scavenging are normally pumped for disposal as mine fill or to tailings disposal facilities for long-term storage.

Froth flotation efficiency is determined by a series of probabilities: those of particle–bubble contact, particle–bubble attachment, transport between the pulp and the froth, and froth collection into the product launder. In a conventional mechanically-agitated cell, the void fraction (i.e. volume occupied by air bubbles) is low (5 to 10 percent) and the bubble size is usually greater than 1 mm. This results in a relatively low interfacial area and a low probability of particle–bubble contact. Consequently, several cells in series are required to increase the particle residence time, thus increasing the probability of particle–bubble contact.

Flotation is normally undertaken in several stages to maximize the recovery of the target mineral or minerals and the concentration of those minerals in the concentrate, while minimizing the energy input.

Flotation Stages

Roughing

The first stage is called *roughing*, which produces a *rougher concentrate*. The objective is to remove the maximum amount of the valuable mineral at as coarse a particle size as practical. The finer an ore is ground, the greater the energy that is required, so it makes sense to fine grind only those particles that need fine grinding. Complete liberation is not required for rougher flotation, only sufficient liberation to release enough gangue from the valuable mineral to get a high recovery.

The primary objective of roughing is to recover as much of the valuable minerals as possible, with less emphasis on the quality of the concentrate produced.

In some concentrators, there may be a *preflotation* step that precedes roughing. This is done when there are some undesirable materials, such as organic carbon, that readily float. They are removed first to avoid them floating during roughing (and thus contaminating the rougher concentrate).

Cleaning

The rougher concentrate is normally subjected to further stages of flotation to reject more of the undesirable minerals that also reported to the froth, in a process known as *cleaning*. The product of cleaning is known as the *cleaner concentrate* or the *final concentrate*.

The objective of cleaning is to produce as high a concentrate grade as possible.

The rougher concentrate is often subject to further grinding (usually called *regrinding*) to get more complete liberation of the valuable minerals.Because it is a smaller mass than that of the original ore, less energy is needed than would be necessary if the whole ore were reground.Regrinding is often undertaken in specialized *regrind mills*, such at the IsaMill™, designed to further reduce the energy consumed during regrinding to finer sizes.

Scavenging

The rougher flotation step is often followed by a *scavenger* flotation step that is applied to the rougher tailings. The objective is to recover any of the target minerals that were not recovered during the initial roughing stage. This might be achieved by changing the flotation conditions to make them more rigorous than the initial roughing, or there might be some secondary grinding to provide further liberation.

The concentrate from the rougher scavengers could be returned to the rougher feed for refloating or sent to special cleaner cells.

Similarly, the cleaning step may also be followed by a scavenging step performed on the cleaner tailings.

Science of Flotation

To be effective on a given ore slurry, the collectors are chosen based upon their selective wetting of the types of particles to be separated. A good collector will adsorb, physically or chemically, with one of the types of particles. This provides the thermodynamic requirement for the particles to bind to the surface of a bubble. The wetting activity of a surfactant on a particle can be quantified by measuring the contact angles that the liquid/bubble interface makes with it. Another important measure for attachment of bubbles to particles is induction time. The induction time is the time required for the particle and bubble to rupture the thin film separating the particle and bubble. This rupturing is achieved by the surface forces between the particle and bubble.

The mechanisms for the bubble-particle attachment is very complex and consists of three steps, collision, attachment and detachment. The collision is achieved by particles being within the collision tube of a bubble and this is affected by the velocity of the bubble and radius of the bubble. The collision tube corresponds to the region in which a particle will collide with the bubble, with the perimeter of the collision tube corresponding to the grazing trajectory.

The attachment of the particle to the bubble is controlled by the induction time of the particle and bubble. The particle and bubble need to bind and this occurs if the time in which the particle and bubble are in contact with each other is larger than the required induction time. This induction time is affected by the fluid viscosity, particle and bubble size and the forces between the particle and bubbles.

The detachment of a particle and bubble occurs when the force exerted by the surface tension is exceeded by shear forces and gravitational forces. These forces are complex and vary within the cell. High shear will be experienced close to the impeller of a mechanical flotation cell and mostly gravitational force in the collection and cleaning zone of a flotation column.

Significant issues of entrainment of fine particles occurs as these particles experience low collision efficiencies as well as sliming and degradation of the particle surfaces. Coarse particles show a low recovery of the valuable mineral due to the low liberation and high detachment efficiencies.

Theory

Selective Adhesion

Froth flotation depends on the selective adhesion of air bubbles to mineral surfaces in a mineral/water slurry. The air bubbles will attach to more hydrophobic particles. The attachment of the bubbles to the surface is determined by the interfacial energies between the solid, liquid, and gas phases. This is determined by the Young-Dupré Equation:

$$\gamma_{lv}\cos\theta = (\gamma_{sv} - \gamma_{sl})$$

where:

- γ_{lv} is the surface energy of the liquid/vapor interface
- γ_{sv} is the surface energy of the solid/vapor interface
- γ_{sl} is the surface energy of the solid/liquid interface,
- θ is the contact angle, the angle formed at the junction between vapor, solid, and liquid phases.

Minerals targeted for separation may be chemically surface-modified with collectors so that they are more hydrophobic. Collectors are a type of surfactant that increase the natural hydrophobicity of the surface, increasing the separability of the hydrophobic and hydrophilic particles. Collectors either chemically bond via chemisorption to the mineral or adsorb onto the surface via physisorption.

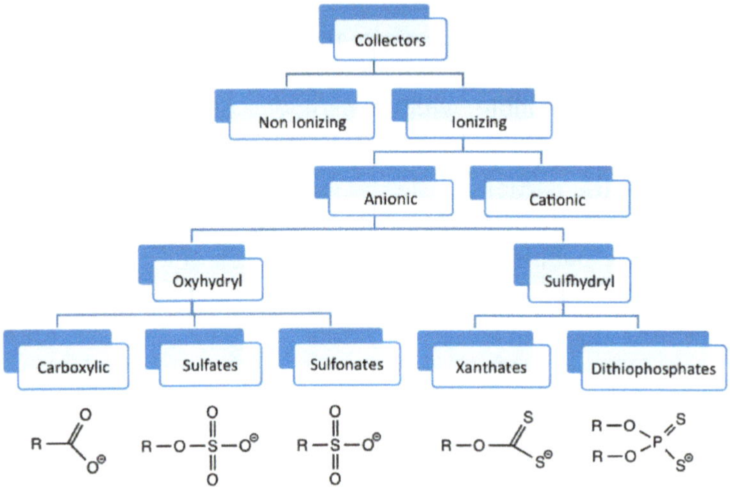

Different types of collectors, or surfactants, used in froth flotation.

IMFs and Surface Forces in Bubble-Particle Interactions

Collision

The collision rates for fine particles (50 - 80 μm) can be accurately modeled, but there is no current theory that accurately models bubble-particle collision for particles as large as 300 μm, which are commonly used in flotation processes.

For fine particles, Stokes law underestimates collision probability while the potential equation based on Surface Charge overestimates collision probability so an intermediate equation is used.

It is important to know the collision rates in the system since this step precedes the adsorption where a three phase system is formed.

Adsorption (Attachment)

The effectiveness of a medium to adsorb to a particle is influenced by the relationship between the surfaces both materials. There are multiple factors that affect the efficiency of adsorption in chemical, thermodynamic, and physical domains. These factors can range from surface energy and polarity to the shape, size, and roughness of the particle. In froth flotation, adsorption is a strong consequence of surface energy, since the small particles have a high surface area to size ratio, resulting in higher energy surfaces to form attractions with adsorbates. The air bubbles must selectively adhere to the desired minerals to elevate them to the surface of the slurry while wetting the other minerals and leaving them in the aqueous slurry medium.

Particles that can be easily wetted by water are called hydrophilic, while particles that are not easily wetted by water are called hydrophobic. Hydrophobic particles have a tendency to form a separate phase in aqueous media. In froth flotation the effectiveness of an air bubble to adhere to a particle is based on how hydrophobic the particle is. Hydrophobic particles have an affinity to air bubbles, leading to adsorption. The bubble-particle combinations are elevated to the froth zone driven by buoyancy forces.

The attachment of the bubbles to the particles is determined by the interfacial energies of between the solid, liquid, and vapor phases, as modeled by the Young/Dupre Equation. The interfacial energies can be based on the natural structure of the materials, or the addition of chemical treatments can improve energy compatibility.

Collectors are the main additives used to improve particle surfaces. They function as surfactants to selectively isolate and aid adsorption between the particles of interest and bubbles rising through the slurry. Common collectors used in flotation are anionic sulfur ligands, which have a bifunctional structure with an ionic portion which shares attraction with metals, and a hydrophobic portion such as a long hydrocarbon tail. These collectors coat a particle's surface with a monolayer of non-polar substance to aid separation from the aqueous phase by decreasing the adsorbed particle solubility in water. The adsorbed ligands can form micelles around the particles and form small-particle colloids improving stability and phase separation further.

Desorption (Detachment)

The adsorption of particles to bubbles is essential to separating the minerals from the slurry, but the minerals must be purified from the additives used in separation, such as the collectors, frothers, and modifiers. The product of the cleaning, or desorption process, is known as the cleaner concentrate. The detachment of a particle and bubble requires adsorption bond cleavage driven by shear forces. Depending on the flotation cell type, shear forces are applied by a variety of mechanical systems. Among the most common are impellers and mixers. Some systems combine the functionalities of these components by placing them at key locations where they can take part in multiple froth flotation mechanisms. Cleaning cells also take advantage of gravitational forces to improve separation efficiency.

Performance Calculations

Relevant Equations

A common quantity used to describe the collection efficiency of a froth flotation process is *flotation recovery* (R). This quantity incorporates the probabilities of collision and attachment of particles to gas flotation bubbles.

$$R = \frac{N_c}{\left(\frac{\pi}{4}\right)(d_p + d_b)^2 Hc}$$

where:

$N_c = PN_c^i$, which is the product of the probability of the particle being collected (P) and the number of possible particle collisions (N_c^i)
d_p is particle diameter
d_b is bubble diameter
H is a specified height within the flotation which the recovery was calculated
c is the particle concentration

The following, are several additional mathematical methods often used to evaluate the effectiveness of froth flotation processes. These equations are more simple than the calculation for *flotation recovery*, as they are based solely on the amounts of inputs and outputs of the processes.

For the following equations:

F is the weight percent of feed
C is the weight percent concentrate
T is the weight percent of tailings
c,t, and f are the Metallurgical assays of the concentrate, tailings, and feed, respectively

Ratio of feed weight to concentrate weight $\frac{F}{C}$ (unitless)

$$\frac{F}{C} = \frac{c-t}{f-t}$$

Percent of metal recovered (X_R) in wt%

$$X_R = 100\left(\frac{c}{f}\right)\left(\frac{f-t}{c-t}\right)$$

Percent of metal lost (X_L) in wt%

$$X_L = 100 - X_R$$

Percent of weight recovered (X_W)_ in wt%

$$X_W = 100\left(\frac{C}{F}\right) = 100\frac{f-t}{c-t}$$

Grade-Recovery Curves

Grade-recovery curves are useful tools in weighing the trade-off of producing a high grade of concentrate while maintaining as low of a recovery rate as possible, two important aspects of froth flotation. These curves are developed empirically based on the individual froth flotation process of a particular plant. As the curves are shifted in the positive x-direction (to the right) and the positive y-direction (upward) the performance of the froth flotation process is regarded as improving. A disadvantage to these curves is that they can only compare the grade-recovery relations of a specific feed grade and feed rate. If a company has a variance of feed grades and rates used (an extremely common occurrence) in their froth flotation process, grade-recovery curves for every pairing of feed grade and recovery rate would have to be constructed in order to provide meaningful information to the plant.

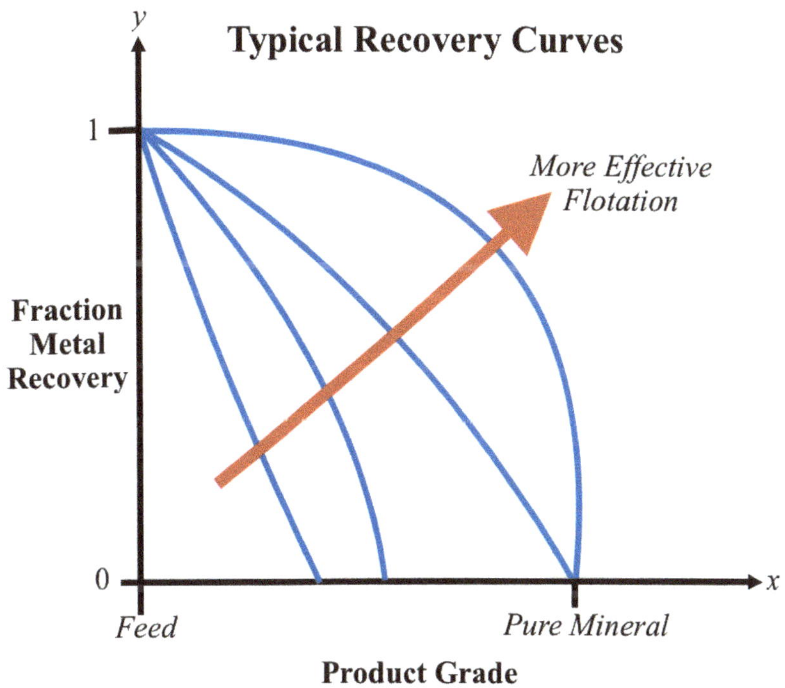

Example grade-recovery relationships seen in froth flotation. Shifts in the curves represent changes in flotation effectiveness.

Flotation Equipment

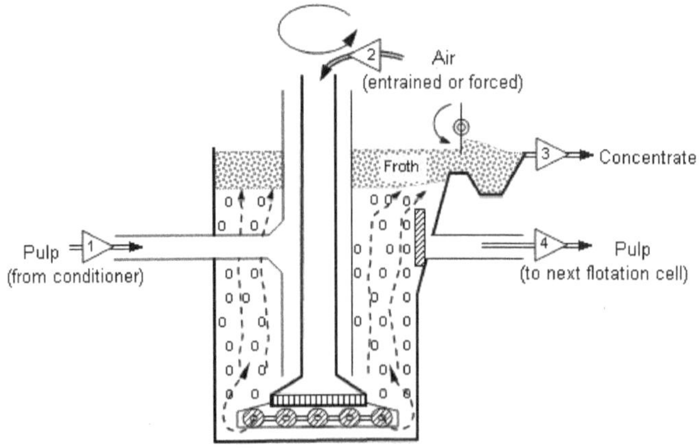

Diagram of froth flotation cell. Numbered triangles show direction of stream flow. A mixture of ore and water called pulp enters the cell from a conditioner, and flows to the bottom of the cell. Air or nitrogen is passed down a vertical impeller where shearing forces break the air stream into small bubbles. The mineral concentrate froth is collected from the top of the cell , while the pulp flows to another cell.

Flotation can be performed in rectangular or cylindrical mechanically agitated cells or tanks, flotation columns, Jameson Cells or deinking flotation machines. Classified by the method of air absorption manner, it is fair to state that two distinct groups of flotation equipments have arisen:pneumatic and mechanical machines. Generally pneumatic machines give a low-grade concentrate and little operating troubles.

Comparison of the sizes of flotation columns and Jameson Cells with similar capacities.

Mechanical cells use a large mixer and diffuser mechanism at the bottom of the mixing tank to introduce air and provide mixing action. **Flotation columns** use air spargers to introduce air at the bottom of a tall column while introducing slurry above. The countercurrent motion of the slurry flowing down and the air flowing up provides mixing action. Mechanical cells generally have a higher throughput rate, but produce material that is of lower quality, while flotation columns generally have a low throughput rate but produce higher quality material.

The Jameson cell uses neither impellers nor spargers, instead combining the slurry with air in a downcomer where high shear creates the turbulent conditions required for bubble particle contacting.

Mechanics of Flotation

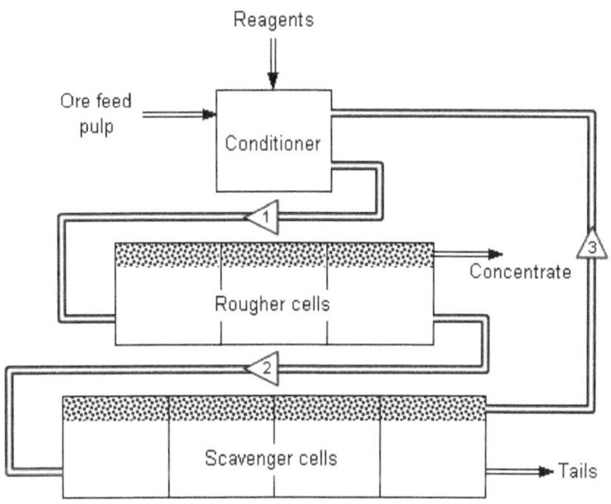

The following steps are followed, following grinding to liberate the mineral particles:

- Reagent conditioning to achieve hydrophobic surface charges on the desired particles
- Collection and upward transport by bubbles in an intimate contact with air or nitrogen
- Formation of a stable froth on the surface of the flotation cell
- Separation of the mineral laden froth from the bath (flotation cell)

Simple flotation circuit for mineral concentration. Numbered triangles show direction of stream flow. Various flotation reagents are added to a mixture of ore and water (called pulp) in a conditioning tank. The flow rate and tank size are designed to give the minerals enough time to be activated. The conditioner pulp is fed to a bank of rougher cells which remove most of the desired minerals as a concentrate. The rougher pulp passes to a bank of scavenger cells where additional reagents may be added. The scavenger cell froth is usually returned to the rougher cells for additional treatment, but in some cases may be sent to special cleaner cells. The scavenger pulp is usually barren enough to be discarded as tails. More complex flotation circuits have several sets of cleaner and re-cleaner cells, and intermediate re-grinding of pulp or concentrate.

Chemicals of Flotation

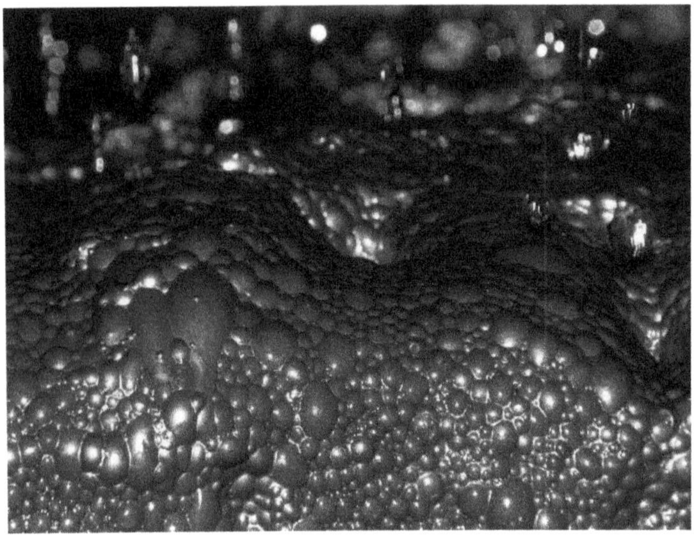

Copper sulfide foam in a froth-flotation cell

Collectors

For many ores (e.g. those of Cu, Mo, W, Ni), the collectors are anionic sulfur ligands. Particularly popular are xanthate salts, including potassium amyl xanthate (PAX), potassium isobutyl xanthate (PIBX), potassium ethyl xanthate (KEX), sodium isobutyl xanthate (SIBX), sodium isopropyl xanthate (SIPX), sodium ethyl xanthate (SEX). Other collectors include related sulfur-based ligands: dithiophosphates, dithiocarbamates. Still other classes of collectors include the thiourea thiocarbanilide. Fatty acids have also been used.

For some minerals (e.g., sylvanite for KCl), fatty amines are used as collectors.

Frothers

A variety of compounds are added to stabilize the foams. These additives include pine oil, various alcohols (methyl isobutyl carbinol (MIBC)), polyglycols, xylenol (cresylic acid).

Modifiers

A variety of other compounds are added to optimize the separation process, these additives are called modifiers. Such species do not interact directly with the mineral, but modify the physical properties of the solution.

- pH modifiers include lime (CaO), Soda ash (Na_2CO_3), Caustic soda (NaOH), sulfuric and hydrochloric acid (H_2SO_4, HCl).
- Anionic modifiers include phosphates, silicates, and carbonates.
- Organic modifiers include the thickeners dextrin, starch, glue, and CMC.

Chemical Compounds for Deinking of Recycled Paper

- pH control: sodium silicate and sodium hydroxide
- Calcium ion source: hard water, lime or calcium chloride
- Collector: fatty acid, fatty acid emulsion, fatty acid soap and/or organo-modified siloxane

Specific ore Applications

Illustrative, the flotation process is used for purification of potassium chloride from sodium chloride and clay minerals. The crushed mineral is suspended in brine in the presence of fatty ammonium salts. Because the ammonium group and K^+ have very similar ionic radii (ca. 0.135, 0.143 nm respectively), the ammonium centers exchange for the surface potassium sites on the particles of KCl, but not on the NaCl particles. The long alkyl chains then confer hydrophobicity, to the particles, which enable them to form foams.

Sulfide ores		
• Copper (see copper extraction)	• Copper-Molybdenum	• Lead-Zinc
• Lead-Zinc-Iron	• Copper-Lead-Zinc-Iron	• Gold-Silver
• Oxide Copper and Lead	• Nickel	• Nickel-Copper
Nonsulfide ores		
• Fluorite	• Tungsten	• Lithium
• Tantalum	• Tin	• Coal

DISSOLVED AIR FLOTATION

Dissolved air flotation (DAF) is a water treatment process that clarifies wastewaters (or other waters) by the removal of suspended matter such as oil or solids. The removal is achieved by dissolving air in the water or wastewater under pressure and then releasing the air at atmospheric pressure in a flotation tank basin. The released air forms tiny bubbles which adhere to the suspended matter causing the suspended matter to float to the surface of the water where it may then be removed by a skimming device.

Dissolved air flotation is very widely used in treating the industrial wastewater effluents from oil refineries, petrochemical and chemical plants, natural gas processing plants, paper mills, general water treatment and similar industrial facilities. A very similar process known as induced gas flotation is also used for wastewater treatment. Froth flotation is commonly used in the processing of mineral ores.

In the oil industry, dissolved gas flotation (DGF) units do not use air as the flotation medium due to the explosion risk. Nitrogen gas is used instead to create the bubbles.

Process Description

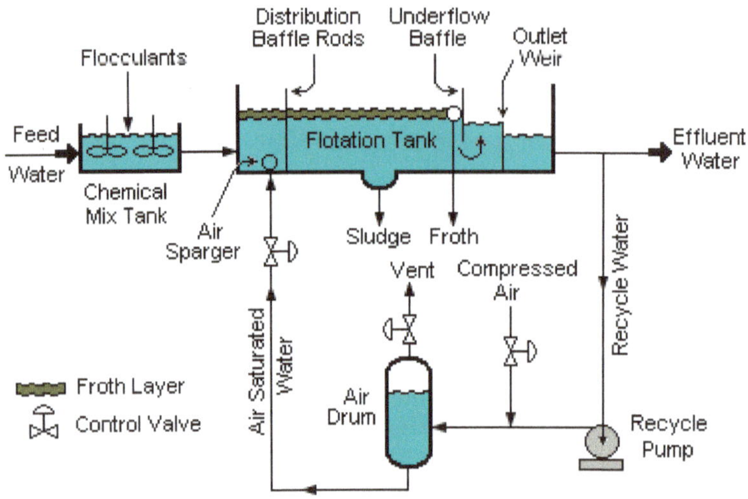

A typical dissolved air flotation unit (DAF)

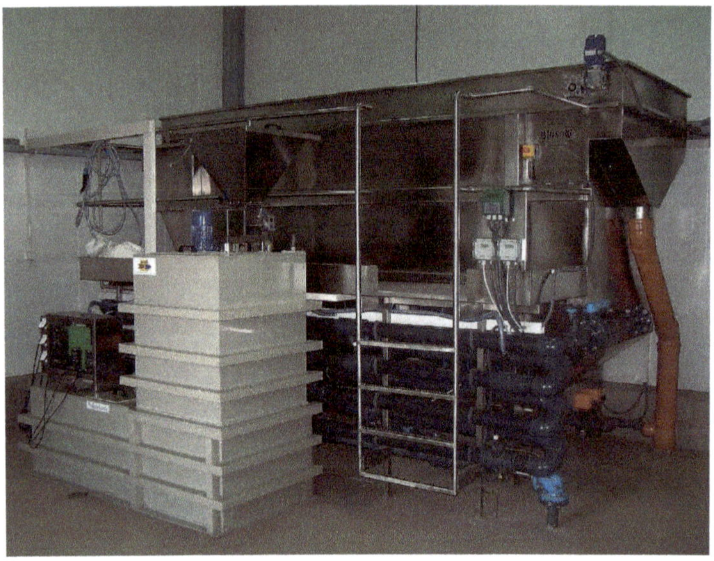

DAF unit with a capacity of 20 m³/h, visible also: flocculant preparation station and pipe flocculator

Modern DAF units using parallel plate technology are quite compact. Picture shows a 225 m^3/h DAF.

The feed water to the DAF float tank is often (but not always) dosed with a coagulant (such as ferric chloride or aluminum sulfate) to coagulate the colloidal particles and/or a flocculant to conglomerate the particles into bigger clusters.

A portion of the clarified effluent water leaving the DAF tank is pumped into a small pressure vessel (called the air drum) into which compressed air is also introduced. This results in saturating the pressurized effluent water with air. The air-saturated water stream is recycled to the front of the float tank and flows through a pressure reduction valve just as it enters the front of the float tank, which results in the air being released in the form of tiny bubbles. Bubbles form at nucleation sites on the surface of the suspended particles, adhering to the particles. As more bubbles form, the lift from the bubbles eventually overcomes the force of gravity. This causes the suspended matter to float to the surface where it forms a froth layer which is then removed by a skimmer. The froth-free water exits the float tank as the clarified effluent from the DAF unit.

Some DAF unit designs utilize parallel plate packing material (e.g. lamellas) to provide more separation surface and therefore to enhance the separation efficiency of the unit.

DAF systems can be categorized as circular (more efficient) and rectangular (more residence time). The former type requires just 3 minutes; an example is a Wockoliver DAF system. The rectangular type requires 20 to 30 minutes; a typical example is a Syskill DAF system. One of the bigger advantages of the circular type is its spiral scoop.

Drinking Water Treatment

Drinking water supplies that are particularly vulnerable to unicellular algal blooms, and supplies with low turbidity and high colour often employ DAF. After coagulation and flocculation processes, water flows to DAF tanks where air diffusers on the tank bottom create fine bubbles that attach to floc resulting in a floating mass of concentrated floc. The floating floc blanket is removed from the surface and clarified water is withdrawn from the bottom of the DAF tank.

DEINKING

Deinking is the industrial process of removing printing ink from paperfibers of recycled paper to make deinked pulp.

The key in the deinking process is the ability to detach ink from the fibers. This is achieved by a combination of mechanical action and chemical means. In Europe the most common process is froth flotation deinking.

Paper is one of the main targets for recycling. A concern about recycling wood pulp paper is that the fibers are degraded with each cycle and after being recycled 4–6 times the fibers become too short and weak to be useful in making paper.

Before the invention of the paper machine in 1799 the most common fibre source was recycled fibres from used textiles, hence the name rag paper. The rags were from hemp, linen and cotton. It was not until the introduction of wood pulp in 1843 that paper production was independent of recycled materials.

Recycling of used paper before the industrialisation of paper production, rag paper was recycled to make low-grade board. A process for removing printing inks from recycled paper was invented by German jurist Justus Claproth in 1774. He practiced together with German paper producer Johann Engelhard Schmid. Today this method is called deinking.

First in the 1950s and 1960s the use of recycled fibres from paper made of wood pulp begun to increase, and was mainly used in packaging paper and paperboard. In the 1950s the froth flotation technique was adapted for deinking recycled paper. Use of recovered paper increased in the 1970s mainly in graphic and hygienic papers, and accelerated in the 1980s. The annual growth in use of recovered paper increased by 6% between 1980 and 1996. The use of virgin fibres only increased 2% in the same period. In 1997 recovered paper production was 42% of the total paper production.

Deinking Process

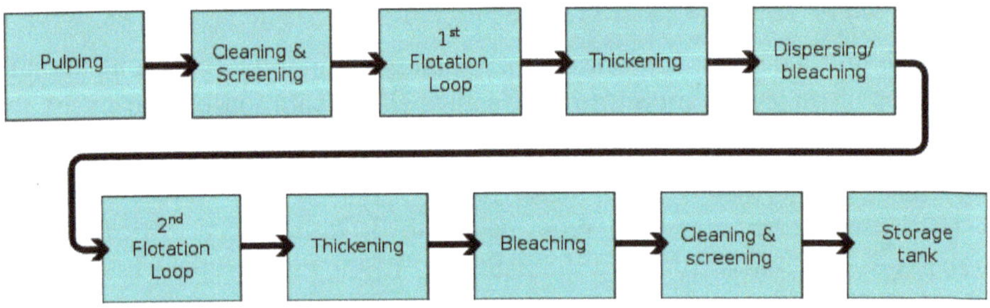

Schematic layout of a deinking plant.

Sorting

Waste paper may contain a mixture of different paper types made of mixtures of different paperfibers. These must be sorted before processed. Broke (paper waste from paper production) is normally used directly in the papermachine.

- Office waste (OW)
- Old magazine papers (OMP)
- Old newsprint (ONP)
- Paperboard
- Corrugated fiberboard

Recycled paper can be used to make paper of the same or lower quality than it was originally. The sorted paper is baled and shipped to a papermill. The pulpmill uses waste paper grade according to the paper quality they want to make.

Debaling

The bales are opened and large foreign objects are sorted out on the conveyor belt to the pulper. Many extraneous materials are readily removed. Twine, strapping, etc. are removed from the hydropulper by a "ragger". Metal straps and staples can be screened out or removed by a magnet. Film-backed pressure-sensitive tape stays intact: the PSA adhesive and the backing are both removed together.

Pulping

Pulpers are either batch, which uses a tub with a high shear rotor, or continuous, using a long, perforated drum. Drum pulpers are very expensive but have the advantage of not breaking up contaminants, thus giving cleaner end product.

The pulper chops the paper to smaller pieces; water and chemicals are added. Normally the pH is adjusted to 8.5 - 10.0. Normal deinking chemicals are:

- pH control: sodium silicate and/or sodium hydroxide
- Bleaching: hydrogen peroxide
- Calcium ion source: hard water, lime or calcium chloride
- Collector: fatty acid, fatty acid emulsion, fatty acid soap and/or organo-modified siloxane

After pulping, the mixture is a slurry. The slurry goes to screening.

Cleaning and Screening

Centrifugal cleaning is spinning the pulp slurry in a cleaner, causing materials that are denser than pulp fibers to move outward and be rejected. Screens, with either slots or holes, are used to remove contaminants that are larger than pulp fibers.

Materials which are more difficult to remove include wax coatings on corrugated cartons and **stickies**, soft rubbery particles which can make deposits and contaminate the recycled paper. Stickies can originate from book bindings, hot melt adhesives, PSA adhesives from paper labels, laminating adhesives of reinforced gummed tapes, etc.

Deinking Stage

In the deinking stage the goal is to release and remove the hydrophobic contaminants from the recycled paper. The contaminants are mostly printing ink and stickies. Several processes are used, most commonly flotation or washing.

Flotation Deinking

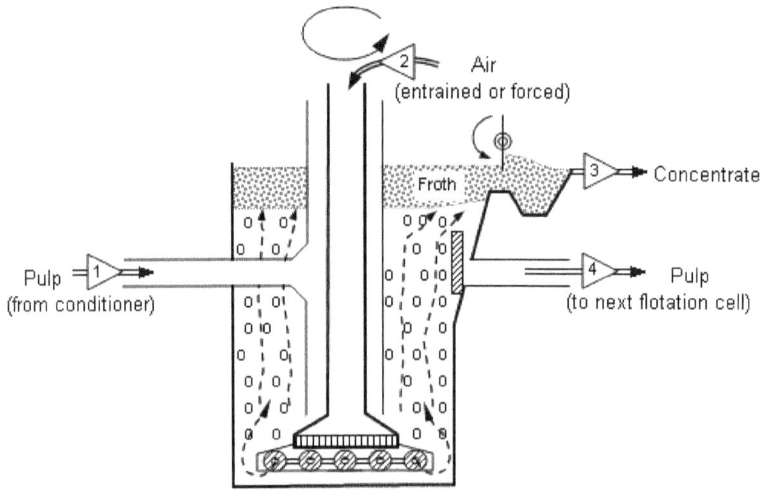

Diagram of a froth flotation cell.

Froth flotation was adapted from the flotation process used in the mining industry in the 1960s. It is the most common deinking process in Europe used to recover recycled paper. Often most of the collector is added to the inlet of the flotation. The process temperatures are normally in the range 45 - 55 °C. Air is blown into the pulp suspension. The collector has affinity both to the ink particles and air bubbles, causing them to attach. The air bubbles lift the ink to the surface and form a thick froth that can be removed. Normally the setup is a two stage system with 3, 4 or 5 flotation cells in series. Flotation deinking is very effective in removing ink particles larger than about 10 μm.

Wash Deinking

Wash deinking consists of a washing stage where dispersants are added to wash out the printing inks. When the pulp slurry is dewatered (thickened), the medium to fine particles are washed out. This process is most useful for removing particles smaller than about 30 μm, like water-based inks, fillers, coating particles, fines and micro stickies. This process is more common when making deinked pulp for tissue. The processing equipment are belt filters, pressure belt filters, disk filters and static filters. This stage is much more efficient than normal washing / dewatering stages.

Combined Washing and Flotation

High quality deinking of office wastes and other printing papers often commonly uses a combination of washing and flotation.

Enzymatic Deinking

This method of deinking uses industrial and/or food grade enzymes in conjunction with flotation deinking to aid in the removal of inks in recycling mills. More efficient removal of ink increases fiber yield, decreases dirt count, and increases paper brightness. Often the use of enzymatic deinking helps mills reduce their bleach usage or use cheaper furnish.

Other Deinking Processes

Dissolved air flotation (DAF) is used by some mills in the deinking stage and will remove some ink and filler (ash); however, it is mainly used to clarify the process water.

Washing / Dewatering

Washing / dewatering (thickening) is a filtration process. Small particles ($< 5\ \mu m$) are removed by passing water through the pulp.

Bleaching

If white paper is desired, bleaching uses peroxides or hydrosulfites to increase the brightness of the pulp. The bleaching methods are similar for mechanical pulp, but the goal is to make the fiber brighter.

Papermaking

The deinked fiber is made into a new paper product in the same way that virgin wood fiber, see papermaking.

Byproducts

The unusable material left over, mainly ink, plastics, filler and short fibers, is called **sludge**. The sludge is buried in a landfill, burned to create energy at the paper mill or used as a fertilizer by local farmers.

Problems

Water based flexographic printing inks with particle sizes below 5 μm and poor solubility in alkaline conditions may cause problems in deinking, especially in the flotation stage. The solution is to use an extra acidic washing stage.

Temperature control is important as this affects the stickiness of stickies.

Additional issues arise when taking into account the number of chemicals potentially present in paper for recycling. Studies have indicated that paper might contain as many as 10,000 different chemicals, fate of which in the deinking process still remains unknown.

FLOCCULATION

Flocculation, in the field of chemistry, is a process wherein colloids come out of suspension in the form of floc or flake; either spontaneously or due to the addition of a clarifying agent. The action differs from precipitation in that, prior to flocculation, colloids are merely suspended in a liquid and not actually dissolved in a solution. In the flocculated system, there is no formation of a cake, since all the flocs are in the suspension.

Coagulation and flocculation are important processes in water treatment with coagulation to destabilize particles through chemical reaction between coagulant and colloids, and flocculation to transport the destabilized particles that will cause collisions with floc.

Term Definition

According to the IUPAC definition, flocculation is "a process of contact and adhesion whereby the particles of a dispersion form larger-size clusters." Flocculation is synonymous with agglomeration and coagulation / coalescence.

Basically, coagulation is a process of addition coagulant to destabilize a stabilized charged particle. Meanwhile, flocculation is a mixing technique that promotes agglomeration and assists in the settling of particles. The most common used coagulant is alum, $Al_2(SO_4)_3{\bullet}14H_2O$.

The chemical reaction involved:

$$Al_2(SO_4)_3{\bullet}14H_2O \rightarrow 2Al(OH)_3(s) + 6H^+ + 3SO_4^{2-} + 8H_2O$$

During flocculation, gentle mixing accelerates the rate of particle collision, and the destabilized particles are further aggregated and enmeshed into larger precipitates. Flocculation is affected by several parameters, including mixing speeds, mixing intensity, and mixing time. The product of the mixing intensity and mixing time is used to describe flocculation processes.

Applications

Surface Chemistry

In colloid chemistry, flocculation refers to the process by which fine particulates are caused to clump together into a floc. The floc may then float to the top of the liquid (creaming), settle to the bottom of the liquid (sedimentation), or be readily filtered from the liquid. Flocculation behavior of soil colloids is closely related to freshwater quality. High dispersibility of soil colloids not only directly causes turbidity of the surrounding water but it also induces eutrophication due to the adsorption of nutritional substances in rivers and lakes.

Physical Chemistry

For emulsions, flocculation describes clustering of individual dispersed droplets together, whereby the individual droplets do not lose their identity.Flocculation is thus the initial step leading to further ageing of the emulsion (droplet coalescence and the ultimate separation of the phases). Flocculation is used in mineral dressing.

Civil Engineering/Earth Sciences

In civil engineering, and in the earth sciences, flocculation is a condition in which clays, polymers or other small charged particles become attached and form a fragile structure, a floc. In dispersed clay slurries, flocculation occurs after mechanical agitation ceases and the dispersed clay platelets spontaneously form flocs because of attractions between negative face charges and positive edge charges.

Biology

Flocculation is used in biotechnology applications in conjunction with microfiltration to improve the efficiency of biological feeds. The addition of synthetic flocculants to the bioreactor can increase the average particle size making microfiltration more efficient. When flocculants are not added, cakes form and accumulate causing low cell viability. Positively charged flocculants work better than negatively charged ones since the cells are generally negatively charged.

Cheese Industry

Flocculation is widely employed to measure the progress of curd formation in the initial stages of cheese making to determine how long the curds must set. The reaction involving the rennet micelles are modeled by Smoluchowski kinetics. During the renneting of milk the micelles can approach one another and flocculate, a process that involves hydrolysis of molecules and macropeptides.

Flocculation is also used during cheese wastewater treatment. Three different coagulants are mainly used:

- $FeSO_4$ (Iron(II) sulfate)
- $Al_2(SO_4)_3$ (Aluminium sulfate)
- $FeCl_3$ (Iron(III) chloride)

Brewing

In the brewing industry flocculation has a different meaning. It is a very important process in fermentation during the production of beer where cells form macroscopic flocs.

These flocs cause the yeast to sediment at the end of the fermentation. Subsequently, the yeast can be collected (cropped) from the top (ale fermentation) or the bottom (lager fermentation) of the fermenter in order to be used for the next fermentation. While it appears similar to sedimentation in colloidal dispersions, the mechanisms are different.

Water Treatment Process

Flocculation and sedimentation are widely employed in the purification of drinking water as well as in sewage treatment, storm-water treatment and treatment of industrial wastewater streams. Typical treatment processes consist of grates, coagulation, flocculation, sedimentation, granular filtration and disinfection.

Jar Test

The purpose of this test is to select types of coagulant (alum) and also to estimate the optimal dose needed in removing the charged particles that occurred in raw water. Jar test is an experiment to understand the processes of coagulation, flocculation and sedimentation (AWWA, 2011).

Jar test apparatus consists of six batch beakers, and equipped with a paddle mixer for each beaker. In a standard practice, jar test involving of rapid mixing, then follow by slow mixing and later with sedimentation process. An example of the flocculation is shown here.

Deflocculation

Deflocculation is the exact opposite of flocculation. Usually in higher pH ranges in addition to low ionic strengths of soil solutions and domination of alkali metal cations the soil colloidal particles can be dispersed. The additive that prevents the colloids from forming flocs is called a deflocculant. According to the Encyclopedic Dictionary of Polymers deflocculation is "a state or condition of a dispersion of a solid in a liquid in which each solid particle remains independent and unassociated with adjacent particles. A deflocculated suspension shows zero or very low yield value".

Deflocculation can be a problem in wastewater treatment plants as it commonly causes sludge settling problems and deterioration of the effluent quality.

FRACTIONAL DISTILLATION

Fractional distillation is the separation of a mixture into its component parts, or fractions, separating chemical compounds by their boiling point by heating them to a temperature at which one or more fractions of the compound will vaporize. It uses distillation to fractionate. Generally the component parts have boiling points that differ by less than 25 °C from each other under a pressure of one atmosphere. If the difference in boiling points is greater than 25 °C, a simple distillation is typically used.

Laboratory Setup

Fractional distillation in a laboratory makes use of common laboratory glassware and apparatuses, typically including a Bunsen burner, a round-bottomed flask and a condenser, as well as the single-purpose fractionating column.

Apparatus

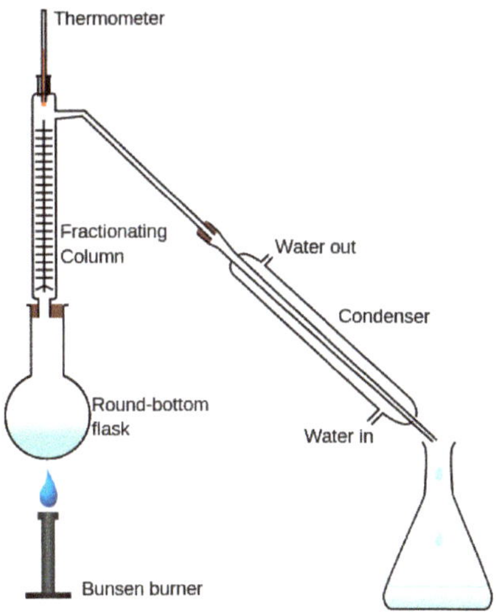

Fractional distillation
An Erlenmeyer flask is used as a receiving flask. Here the distillation head and fractionating column are combined in one piece.

- heat source, such as a hot plate with a bath, and ideally with a magnetic stirrer.
- distilling flask, typically a round-bottom flask
- receiving flask, often also a round-bottom flask
- fractionating column
- distillation head
- thermometer and adapter if needed
- condenser, such as a Liebig condenser or Allihn condenser
- vacuum adapter (not used in image to the right)
- boiling chips, also known as anti-bumping granules
- Standard laboratory glassware with ground glass joints, e.g. quickfit apparatus.

Discussion

As an example consider the distillation of a mixture of water and ethanol. Ethanol boils at 78.4 °C while water boils at 100 °C. So, by heating the mixture, the most volatile component (ethanol) will concentrate to a greater degree in the vapor leaving the liquid. Some mixtures form azeotropes, where the mixture boils at a lower temperature than either component. In this example, a mixture of 96% ethanol and 4% water boils at 78.2 °C; the mixture is more volatile than pure ethanol. For this reason, ethanol cannot be completely purified by direct fractional distillation of ethanol-water mixtures.

The apparatus is assembled as in the diagram. (The diagram represents a batch apparatus as opposed to a continuous apparatus.) The mixture is put into the round bottomed flask along with a few anti-bumping granules (or a Teflon coated magnetic stirrer bar if using magnetic stirring), and the fractionating column is fitted into the top. The fractional distillation column is set up with the heat source at the bottom on the still pot. As the distance from the stillpot increases, a temperature gradient is formed in the column; it is coolest at the top and hottest at the bottom. As the mixed vapor ascends the temperature gradient, some of the vapor condenses and revaporizes along the temperature gradient. Each time the vapor condenses and vaporizes, the composition of the more volatile component in the vapor increases. This distills the vapor along the length of the column, and eventually the vapor is composed solely of the more volatile component (or an azeotrope). The vapor condenses on the glass platforms, known as trays, inside the column, and runs back down into the liquid below, refluxing distillate. The efficiency in terms of the amount of heating and time required to get fractionation can be improved by insulating the outside of the column in an insulator such as wool, aluminium foil or preferably a vacuum jacket. The hottest tray is at the bottom and the coolest is at the top. At steady state conditions, the vapor and liquid on each tray are at *equilibrium*. The most volatile component of the mixture exits as a gas at the top of the column. The vapor at the top of the column then passes into the condenser, which cools it down until it liquefies. The separation is more pure with the addition of more trays (to a practical limitation of heat, flow, etc.) Initially, the condensate will be close to the azeotropic composition, but when much of the ethanol has been drawn off, the condensate becomes gradually richer in water. The process continues until all the ethanol boils out of the mixture. This point can be recognized by the sharp rise in temperature shown on the thermometer.

The above explanation reflects the theoretical way fractionation works. Normal laboratory fractionation columns will be simple glass tubes (often vacuum-jacketed, and sometimes internally silvered) filled with a packing, often small glass helices of 4 to 7 mm diameter. Such a column can be calibrated by the distillation of a known mixture system to quantify the column in terms of number of theoretical trays. To improve fractionation the apparatus is set up to return condensate to the column by the use of some sort of reflux splitter (reflux wire, gago, Magnetic swinging bucket, etc.) - a typical careful fractionation would employ a reflux ratio of around 4:1 (4 parts returned condensate to 1 part condensate take off).

In laboratory distillation, several types of condensers are commonly found. The Liebig condenser is simply a straight tube within a water jacket, and is the simplest (and relatively least expensive) form of condenser. The Graham condenser is a spiral tube within a water jacket, and the Allihn condenser has a series of large and small constrictions on the inside tube, each increasing the surface area upon which the vapor constituents may condense.

Alternate set-ups may use a multi–outlet distillation receiver flask (referred to as a "cow" or "pig") to connect three or four receiving flasks to the condenser. By turning the cow or pig, the distillates can be channeled into any chosen receiver. Because the receiver does not have to be removed and replaced during the distillation process, this type of apparatus is useful when distilling under an inert atmosphere for air-sensitive chemicals or at reduced pressure. A Perkin triangle is an alternative apparatus often used in these situations because it allows isolation of the receiver from the rest of the system, but does require removing and reattaching a single receiver for each fraction.

Vacuum distillation systems operate at reduced pressure, thereby lowering the boiling points of the materials. Anti-bumping granules, however, become ineffective at reduced pressures.

Industrial Distillation

Typical industrial fractional distillation columns

Fractional distillation is the most common form of separation technology used in petroleum refineries, petrochemical and chemical plants, natural gas processing and cryogenic air separation plants. In most cases, the distillation is operated at a continuous steady state. New feed is always being added to the distillation column and products are always being removed. Unless the process is disturbed due to changes in feed, heat, ambient temperature, or condensing, the amount of feed being added and the amount of product being removed are normally equal. This is known as continuous, steady-state fractional distillation.

Industrial distillation is typically performed in large, vertical cylindrical columns known as "distillation or fractionation towers" or "distillation columns" with diameters ranging from about 65 centimeters to 6 meters and heights ranging from about 6 meters to 60 meters or more. The distillation towers have liquid outlets at intervals up the column which allow for the withdrawal of different fractions or products having different boiling points or boiling ranges. By increasing the temperature of the product inside the columns, the different hydrocarbons are separated. The "lightest" products (those with the lowest boiling point) exit from the top of the columns and the "heaviest" products (those with the highest boiling point) exit from the bottom of the column.

For example, fractional distillation is used in oil refineries to separate crude oil into useful substances (or fractions) having different hydrocarbons of different boiling points. The crude oil fractions with higher boiling points:

- have more carbon atoms
- have higher molecular weights
- are less branched chain alkanes
- are darker in color
- are more viscous
- are more difficult to ignite and to burn

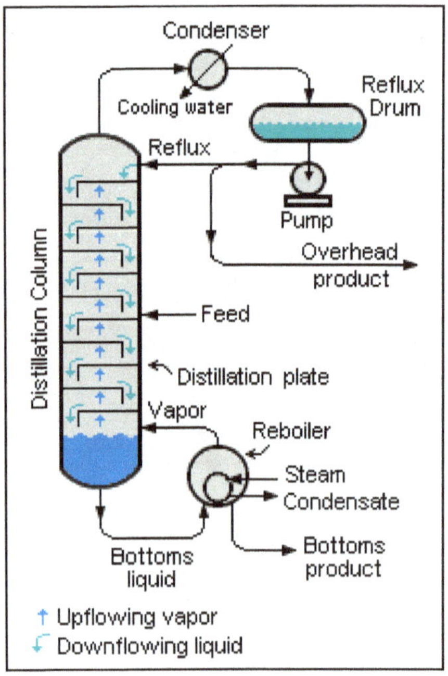

Diagram of a typical industrial distillation tower

Large-scale industrial towers use reflux to achieve a more complete separation of products. Reflux refers to the portion of the condensed overhead liquid product from a distillation or fractionation tower that is returned to the upper part of the tower as shown in the schematic diagram of a typical, large-scale industrial distillation tower. Inside the tower, the reflux liquid flowing downwards provides the cooling needed to condense the vapors flowing upwards, thereby increasing the effectiveness of the distillation tower. The more reflux is provided for a given number of theoretical plates, the better the tower's separation of lower boiling materials from higher boiling materials. Alternatively, the more reflux provided for a given desired separation, the fewer theoretical plates are required.

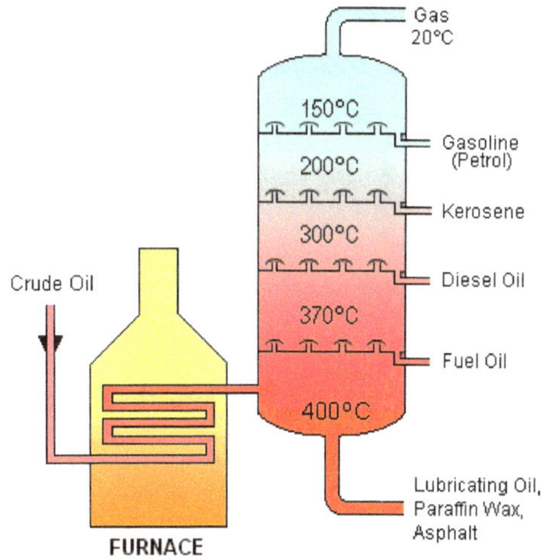

Crude oil is separated into fractions by fractional distillation. The fractions at the top of the fractionating column have lower boiling points than the fractions at the bottom. All of the fractions are processed further in other refining units.

Fractional distillation is also used in air separation, producing liquid oxygen, liquid nitrogen, and highly concentrated argon. Distillation of chlorosilanes also enable the production of high-purity silicon for use as a semiconductor.

In industrial uses, sometimes a packing material is used in the column instead of trays, especially when low pressure drops across the column are required, as when operating under vacuum. This packing material can either be random dumped packing (1-3 wide) such as Raschig rings or structured sheet metal. Typical manufacturers are Koch, Sulzer and other companies. Liquids tend to wet the surface of the packing and the vapors pass across this wetted surface, where mass transfer takes place. Unlike conventional tray distillation in which every tray represents a separate point of vapor liquid equilibrium the vapor liquid equilibrium curve in a packed column is continuous. However, when modeling packed columns it is useful to compute a number of theoretical plates to denote the separation efficiency of the packed column with respect to more traditional trays. Differently shaped packings have different surface areas and void space between packings. Both of these factors affect packing performance.

Design of Industrial Distillation Columns

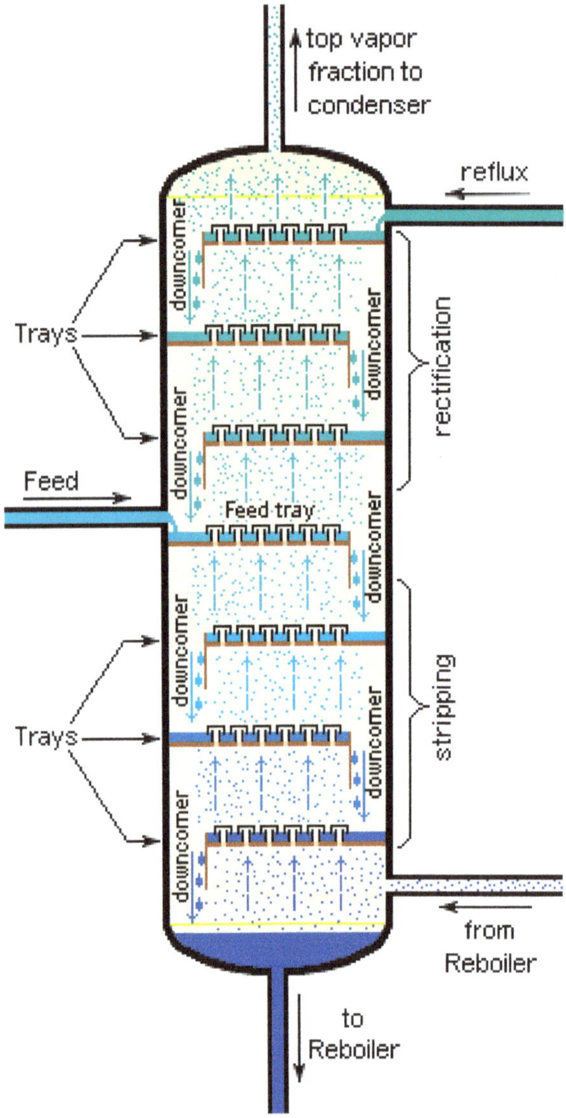

Chemical engineering schematic of typical bubble-cap trays in a distillation tower

Design and operation of a distillation column depends on the feed and desired products. Given a simple, binary component feed, analytical methods such as the McCabe–Thiele method or the Fenske equation can be used. For a multi-component feed, simulation models are used both for design and operation.

Moreover, the efficiencies of the vapor–liquid contact devices (referred to as *plates* or *trays*) used in distillation columns are typically lower than that of a theoretical 100% efficient equilibrium stage. Hence, a distillation column needs more plates than the number of theoretical vapor–liquid equilibrium stages.

Reflux refers to the portion of the condensed overhead product that is returned to the tower. The reflux flowing downwards provides the cooling required for condensing the vapours flowing upwards. The reflux ratio, which is the ratio of the (internal) reflux to the overhead product, is conversely related to the theoretical number of stages required for efficient separation of the distillation products. Fractional distillation towers or columns are designed to achieve the required separation efficiently. The design of fractionation columns is normally made in two steps; a process design, followed by a mechanical design. The purpose of the process design is to calculate the number of required theoretical stages and stream flows including the reflux ratio, heat reflux and other heat duties. The purpose of the mechanical design, on the other hand, is to select the tower internals, column diameter and height. In most cases, the mechanical design of fractionation towers is not straightforward. For the efficient selection of tower internals and the accurate calculation of column height and diameter, many factors must be taken into account. Some of the factors involved in design calculations include feed load size and properties and the type of distillation column used.

The two major types of distillation columns used are tray and packing columns. Packing columns are normally used for smaller towers and loads that are corrosive or temperature sensitive or for vacuum service where pressure drop is important. Tray columns, on the other hand, are used for larger columns with high liquid loads. They first appeared on the scene in the 1820s. In most oil refinery operations, tray columns are mainly used for the separation of petroleum fractions at different stages of oil refining.

In the oil refining industry, the design and operation of fractionation towers is still largely accomplished on an empirical basis. The calculations involved in the design of petroleum fractionation columns require in the usual practice the use of numerable charts, tables and complex empirical equations. In recent years, however, a considerable amount of work has been done to develop efficient and reliable computer-aided design procedures for fractional distillation.

FRACTIONAL FREEZING

Fractional freezing is a process used in process engineering and chemistry to separate substances with different melting points. It can be done by partial melting of a solid, for example in zone refining of silicon or metals, or by partial crystallization of a liquid, as in freeze distillation, also called normal freezing or progressive freezing. The initial sample is thus fractionated (separated into fractions).

Partial crystallization can also be achieved by adding a dilute solvent to the mixture, and cooling and concentrating the mixture by evaporating the solvent, a process called solution crystallization. Fractional freezing is generally used to produce ultra-pure solids, or to concentrate heat-sensitive liquids.

Freeze Distillation

Eisbock beer (12% alcohol) created via freeze distillation of doppelbock beer. Barrels of beer were originally left outdoors to partially freeze, then the ice removed.

Freeze distillation is a misnomer, since it is not distillation but rather a process of enriching a solution by partially freezing it and removing frozen material that is poorer in the dissolved material than is the liquid portion left behind. Such enrichment parallels enrichment by true distillation, where the evaporated and re-condensed portion is richer than the liquid portion left behind.

The detailed situation is the subject of thermodynamics, a subdivision of physics of importance to chemistry. Without resorting to mathematics, the following can be said for a mixture of water and alcohol:

- Freezing in this scenario begins at a temperature significantly below 0 °C.
- The first material to freeze is not the water, but a dilute solution of alcohol in water.
- The liquid left behind is richer in alcohol, and as a consequence, further freezing would take place at progressively lower temperatures. The frozen material, while always poorer in alcohol than the (increasingly rich) liquid, becomes progressively richer in alcohol.
- Further stages of removing frozen material and waiting for more freezing will come to naught once the liquid uniformly cools to the temperature of whatever is cooling it.

- If progressively colder temperatures are available,
    - the frozen material will contain progressively larger concentrations of alcohol, and
    - the fraction of the original alcohol removed with the solid material will increase.
- In practice, unless the removal of solid material carries away liquid, the degree of concentration will depend on the final temperature rather than on the number of cycles of removing solid material and chilling.
- Thermodynamics gives fair assurance, even without more information about alcohol and water than that they freely dissolve in each other, that
    - even if temperatures *somewhat below* the freezing point of ethyl alcohol are achieved, there will still be alcohol and water mixed as a liquid, and
    - at some still lower temperature, the remaining alcohol-and-water solution will freeze without an alcohol-poor solid being separable.

Fractional freezing of grapes to concentrate grape sugar for making ice wine.

The best-known freeze-distilled beverages are applejack and ice beer. Ice wine is the result of a similar process, but in this case, the freezing happens *before* the fermentation, and thus it is sugar, not alcohol, that gets concentrated. For an in depth discussion of the physics and chemistry, see eutectic point.

Purification of Solids

When a pure solid is desired, two possible situations can occur. If the contaminant is soluble in the desired solid, a multiple stage fractional freezing is required, analogous to multistage distillation. If, however, a eutectic system forms (analogous to an azeotrope in distillation), a very pure solid can be recovered, as long as the liquid is not at its eutectic composition (in which case a mixed solid forms, which can be hard to separate) or above its eutectic composition (in which case the undesired solid forms).

Concentration of Liquids

When the requirement is to concentrate a liquid phase, fractional freezing can be useful due to its simplicity. Fractional freezing is also used in the production of fruit juice concentrates and other heat-sensitive liquids, as it does not involve heating the liquid (as happens during evaporation).

Desalination

Fractional freezing can be used to desalinate sea water. In a process that naturally occurs with sea ice, frozen salt water, when partially melted, leaves behind ice that is of a much lower salt content. Because sodium chloride lowers the melting point of water, the salt in sea water tends to be forced out of pure water while freezing. Likewise, the frozen water with the highest concentration of salt melts first. Either method decreases the salinity of the frozen water left over, and with multiple runs can be drinkable.

Alcoholic Beverages

Fractional freezing can be used as a simple method to increase the alcohol concentration in fermented alcoholic beverages, a process sometimes called freeze distillation. Examples are applejack, made from hard cider, and ice beer. In practice, while not able to produce an alcohol concentration comparable to distillation, this technique can achieve some concentration with far less effort than any practical distillation apparatus would require. The danger of freeze distillation of alcoholic beverages, is that unlike heat distillation, where the methanol and other impurities can be separated from the finished product, freeze distillation does not remove them. Thus the ratio of impurities may be increased compared to the total volume of the beverage. This concentration may cause side effects to the drinker, leading to intense hangovers and a condition known as "apple palsy" (although this term has also simply been used to refer to intoxication, especially from applejack.)

Alternative Fuels

Fractional freezing is commonly used as a simple method to reduce the gel point of biodiesel and other alternative diesel fuels, whereby esters of higher gel point are removed from esters of lower gel point through cold filtering, or other methods to reduce the subsequent alternative fuel gel point of the fuel blend. This process employs fuel stratification whereby components in the fuel blend develop a higher specific gravity as they approach their respective gel points and thus sink to the bottom of the container, where they can be removed.

REFERENCES

1. G J Jameson, “Flotation cell development,” in: The AusIMM Annual Conference, Broken Hill, New South Wales, 17–21 May 1992 (The Australasian Institute of Mining and Metallurgy: Melbourne, 1992), 25–31.
2. “Wales - The birthplace of Flotation”. Retrieved 2010-01-13.
3. Rickard, Thomas A. (1922). Interviews with Mining Engineers. San Francisco: Mining and Scientific Press. pp. 119–131.
4. Osborne, Graeme (1981). “Guillaume Daniel Delprat”. Australian Dictionary of Biography. Canberra: Australian National University. Retrieved 7 June 2012.
5. “Historical Note”. Minerals Separation Ltd. Retrieved 2007-12-30.
6. Callow; 1916
7. Rickard, Thomas A. (1922). Interviews with Mining Engineers. San Francisco: Mining and Scientific Press. p. 142.
8. A detailed description of the history of flotation and this process can be found in Callows “Notes on Flotation” found in the Transactions of the American Institute of Mining Engineers; Vol 53-54, originally presented in New York in February 1916.
9. Wills, B A; Atkinson, K (1991). “The development of minerals engineering in the 20th Century”. Minerals Engineering. 4: 643–652. doi:10.1016/0892-6875(91)90054-y.
10. Beychok, Milton R. (1967). Aqueous Wastes from Petroleum and Petrochemical Plants (1st ed.). John Wiley & Sons Ltd. LCCN 67019834.
11. Voith EcoCell flotation plant http://www.voithpaper.com/applications/productsearch/files/594_VPR-PB-07-0001-GB-07.pdf
12. D N Nihill, C M Stewart and P Bowen, “The McArthur River mine—the first years of operation,” in: AusIMM ’98 – The Mining Cycle, Mount Isa, 19–23 April 1998 (The Australasian Institute of Mining and Metallurgy: Melbourne, 1998), 73–82.
13. E V Manlapig, C Green, J W Parkinson and A S Murphy, “The technology and economic incentives for recovering coal from tailings impoundments,” SME Annual Meeting, Denver, Colorado, 26–28 February 2001, Preprint 01-70 (Society of Mining, Metallurgy and Exploration: Littleton, Colorado, 2001).
14. B W Atkinson, C J Conway and G J Jameson, “Fundamentals of Jameson Cell operation including size–yield response,” in: Sixth Australian Coal Preparation Conference, Mackay, Queensland, 6–9 September 1993 (The Australasian Institute of Mining and Metallurgy: Melbourne, 1993).
15. B W Atkinson, C J Conway and G J Jameson, “High-efficiency flotation of coarse and fine coal,” in: High-efficiency Coal Preparation: An International Symposium, (Society of Mining, Metallurgy and Exploration: Littleton, Colorado, 1995).

16. J Pease, “Increasing the energy efficiency of grinding,” Presented at: Crushing and Grinding, Brisbane, September 2007. Accessed 24 May 2013.
17. T Smith, D Lin, B Lacouture and G Anderson, “Removal of organic carbon with a Jameson Cell at Red Dog Mine,” in: Proceedings of the 40th Annual Meeting of the Canadian Mineral Processors, Ottawa, Ontario, 22–24 January 2008. Accessed 6 June 2013.
18. Kawatra, S.K. “Flotation Fundamentals” (PDF). MTU Chemistry. Retrieved 8 June 2015.
19. Nguyen, Anh V (12 June 1996). “On modelling of bubble–particle attachment probability in flotation”. International Journal of Mineral Processing. 53: 225–249. doi:10.1016/S0301-7516(97)00073-2.
20. Shahbazi, B. “Bubble–particle collision and attachment probability on fine particles flotation”. Chemical Engineering and Processing: Process Intensification. 49: 622–627. doi:10.1016/j.cep.2010.04.009.
21. Kawatra, S.K. “Flotation Fundamentals” (PDF). MTU Chemistry. Retrieved 8 June 2015.
22. Kawatra, S. K. “Froth Flotation – Fundamental Principles.” (n.d.): n. pag. Web.
23. Neethling, S.j., and J.j. Cilliers. “Grade-recovery Curves: A New Approach for Analysis of and Predicting from Plant Data.” Minerals Engineering 36-38 (2012): 105-10. Web.
24. WO 011717, Nellesen, Bernhard & Christina Northfleet, “METHOD OF DEINKING”, published 05.02.2004
25. Elizabeth R. Burkhardt “Potassium and Potassium Alloys” in Ullmann’s Encyclopedia of Industrial Chemistry, Wiley-VCH, 2006. doi:10.1002/14356007.a22_031.pub2
26. Sebenik, Roger F. et al. (2005) “Molybdenum and Molybdenum Compounds” in Ullmann’s Encyclopedia of Chemical Technology. Wiley-VCH, Weinheim. doi: 10.1002/14356007.a16_655
27. Beychok, Milton R. (1967). Aqueous Wastes from Petroleum and Petrochemical Plants (1st ed.). John Wiley & Sons. LCCN 67019834.
28. Lawrence K. Wang; Yung-Tse Hung; Howard H. Lo; Constantine Yapijakis (2004). Handbook of Industrial and Hazardous Wastes Treatment (2nd ed.). CRC Press. ISBN 0-8247-4114-5.
29. Kiuru, H.; Vahala, R., eds. (2000). “International conference on DAF in water and waste water treatment No. 4, Helsinki, Finland”. IWA Publishing, London. ISBN 1-900222-81-7. |contribution= ignored (help)
30. Paolo Scardina; Marc Edwards (February 11, 2000). “Water Treatment: Fundamentals and Practical Implications of Bubble Formation” (PDF): 3. Retrieved 2015-05-02.
31. Edzwald, James K., ed. (2011). Water Quality and Treatment. 6th Edition. New York:McGraw-Hill. pp. 9.46. ISBN 978-0-07-163011-5

32. "Paper Recycling Information Sheet". Waste Online. Retrieved October 20, 2007.

33. Göttsching, Lothar; Pakarinen, Heikki (2000). "1". Recycled Fiber and Deinking. Papermaking Science and Technology. 7. Finland: Fapet Oy. pp. 12–14. ISBN 952-5216-07-1.

34. Sixta, Herbert, ed. (2006). Handbook of Pulp. 2. Germany: Wiley-VCH. p. 1174. ISBN 3-527-30999-3.

35. Jensen, Timothy (April 1999). "Packaging Tapes: To Recycle of Not". Adhesives and Sealants Council. Archived from the original on 2007-11-09. Retrieved 2007-11-06.

36. WO 011717, Nellesen, Bernhard & Christina Northfleet, "METHOD OF DEINKING", published 05.02.2004, issued 2004

37. "Recycling Compatible Adhesives Standards". Tag and Label Manufacturers Institute. 2007. Archived from the original on 2007-11-09. Retrieved 2007-11-06.

38. "Voluntary Standard for Repulping and Recycling Corugated Fiberboard" (PDF). Corrugated Packaging Alliance. 2005. Archived from the original (PDF) on 2007-12-03. Retrieved 2007-11-06.

39. Seiter, Pikulin, (October 1998). "Environmentally benign USPS stamps" (PDF). TAPPI Pulping Conference. Retrieved 2007-11-08.

40. Voith EcoCell flotation plant http://www.voithpaper.com/applications/productsearch/files/594_VPR-PB-07-0001-GB-07.pdf

41. Pivnenko, Kostyantyn; Eriksson, Eva; Astrup, Thomas F. "Waste paper for recycling: Overview and identification of potentially critical substances". Waste Management. doi:10.1016/j.wasman.2015.02.028.

42. "Terminology of polymers and polymerization processes in dispersed systems (IUPAC Recommendations 2011)" (PDF). Pure and Applied Chemistry. 83 (12): 2229–2259. 2011. doi:10.1351/PAC-REC-10-06-03.

43. Richard G. Jones; Edward S. Wilks; W. Val Metanomski; Jaroslav Kahovec; Michael Hess; Robert Stepto; Tatsuki Kitayama, eds. (2009). Compendium of Polymer Terminology and Nomenclature (IUPAC Recommendations 2008) "The Purple Book" (2nd ed.). RSC Publishing. ISBN 978-0-85404-491-7.

44. IUPAC, Compendium of Chemical Terminology, 2nd ed. (the "Gold Book") (1997). Online corrected version: (2006–) "flocculation".

45. Hubbard, Arthur T. (2004). Encyclopedia of Surface and Colloid Science. CRC Press. p. 4230. ISBN 0-8247-0759-1. Retrieved 2007-11-13.

46. Adamson A.W. and Gast A.P. (1997) "Physical Chemistry of Surfaces", John Wiley and Sons.

47. Investigation of laws of selective flocculation of coals with synthetic latexes / P. V. Sergeev, V. S. Biletskyy // ICCS'97. 7–12 September 1997, Essen, Germany. V. 1. pp. 503–506.

48. Han, Binbing; Akeprathumchai, S.; Wickramasinghe, S. R.; Qian, X. (2003-07-01). "Flocculation of biological cells: Experiment vs. theory". AIChE Journal. 49 (7): 1687–1701. doi:10.1002/aic.690490709. ISSN 1547-5905.

49. Fox, Patrick F. (1999). Cheese Volume 1: Chemistry, Physics, and Microbiology (2nd ed.). Gaithersburg, Maryland: Aspen Publishers. pp. 144–150. ISBN 978-0-8342-1378-4.

50. Journal of Scientific and Industrial Research. 57: 680–681. 1998. Missing or empty |title= (help)

51. Fox, Patrick F. (2004). Cheese - Chemistry, Physics and Microbiology (3rd Edition). Elsevier. p. 72. ISBN 978-0-12-263653-0.

52. Rivas, Javier; Prazeres, Ana R.; Carvalho, Fatima; Beltrán, Fernando (2010-07-14). "Treatment of Cheese Whey Wastewater: Combined Coagulation−Flocculation and Aerobic Biodegradation". Journal of Agricultural and Food Chemistry. 58 (13): 7871–7877. doi:10.1021/jf100602j. ISSN 0021-8561.

53. Ginovart, M.; López, D.; Giró, A.; Silbert, M. (2006-01-01). "Flocculation in brewing yeasts: A computer simulation study". Biosystems. 83 (1): 51–55. doi:10.1016/j.biosystems.2005.09.002.

54. Beverly, Richard P (2014-04-17). "Water Treatment Process Monitoring and Evaluation". Knovel. American Water Works Association (AWWA). Retrieved October 14, 2015.

55. "Effect of Adding Flocculant to Slurry".

56. Gooch, Dr Jan W., ed. (2007-01-01). Deflocculation. Springer New York. pp. 265–265. doi:10.1007/978-0-387-30160-0_3313. ISBN 978-0-387-31021-3.

57. Perry, Robert; Don Green (2007). Perry's Chemical Engineers' Handbook. McGraw-Hill International Editions. pp. 17–3 to 17–4. ISBN 0-07-142294-3.

58. Janik, Erika (2011). Apple a global history. London: Reaktion Books. ISBN 9781861899583. Retrieved 3 October 2014.

59. Kaufman, Martin (1979). The University of Vermont College of Medicine. Burlington, Vt.: University of Vermont College of Medicine. p. 12. ISBN 9780874511482. Retrieved 3 October 2014.

60. Nordegren, Thomas (2002). The A-Z encyclopedia of alcohol and drug abuse. Parkland, Fla.: Brown Walker Press. p. 78. ISBN 9781581124040. Retrieved 3 October 2014.

CHAPTER 8

Natural-Gas Processing

Chapter Contents

Patented gas processing technology and proprietary know-how developed by CB&I are used in more than 200 natural gas plants around the world. Many of the innovations we have developed, like the use of plate-fin exchangers and packings in cryogenic columns, remain today as standard designs in the industry. We have a broad technology portfolio—including gas conditioning and treating, fractionation, deep ethane extraction and LPG recovery, and liquid product treating—enabling us to offer solutions across the natural gas value chain.

INTRODUCTION

Natural-gas processing is a complex industrial process designed to clean raw natural gas by separating impurities and various non-methane hydrocarbons and fluids to produce what is known as *pipeline quality* dry natural gas.

Natural-gas processing begins at the well head. The composition of the raw natural gas extracted from producing wells depends on the type, depth, and location of the underground deposit and the geology of the area. Oil and natural gas are often found together in the same reservoir. The natural gas produced from oil wells is generally classified as *associated-dissolved*, meaning that the natural gas is associated with or dissolved in crude oil. Natural gas production absent any association with crude oil is classified as "non-associated." In 2009, 89 percent of U.S. wellhead production of natural gas was non-associated.

Natural-gas processing plants purify raw natural gas by removing common contaminants such as water, carbon dioxide (CO_2) and hydrogen sulfide (H_2S). Some of the substances which contaminate natural gas have economic value and are further processed or sold. A fully operational plant delivers pipeline-quality dry natural gas that can be used as fuel by residential, commercial and industrial consumers.

A natural-gas processing plant

Types of raw-Natural-Gas Wells

Raw natural gas comes primarily from any one of three types of wells: crude oil wells, gas wells, and condensate wells.

Natural gas that comes from crude oil wells is typically called *associated gas*. This gas can have existed as a gas cap above the crude oil in the underground formation, or could have been dissolved in the crude oil.

Natural gas from gas wells and from condensate wells, in which there is little or no crude oil, is called *non-associated gas*. Gas wells typically produce only raw natural gas, while condensate wells produce raw natural gas along with other low molecular weight hydrocarbons. Those that are liquid at ambient conditions (i.e., pentane and heavier) are called *natural gas condensate* (sometimes also called *natural gasoline* or simply *condensate*).

Natural gas is called *sweet gas* when relatively free of hydrogen sulfide; gas that does contain hydrogen sulfide is called *sour gas*. Natural gas, or any other gas mixture, containing significant quantities of hydrogen sulfide, carbon dioxide or similar acidic gases, is called *acid gas*

Raw natural gas can also come from methane deposits in the pores of coal seams, and especially in a more concentrated state of adsorption onto the surface of the coal itself. Such gas is referred to as *coalbed gas* or *coalbed methane* (*coal seam gas* in Australia). Coalbed gas has become an important source of energy in recent decades.

Contaminants in Raw Natural Gas

Raw natural gas typically consists primarily of methane (CH_4), the shortest and lightest hydrocarbon molecule. It also contains varying amounts of:

- Heavier gaseous hydrocarbons: ethane (C_2H_6), propane (C_3H_8), normal butane (n-C_4H_{10}), isobutane (i-C_4H_{10}), pentanes and even higher molecular weight hydrocarbons. When processed and purified into finished by-products, all of these are collectively referred to as Natural Gas Liquids or **NGL**.
- Acid gases: carbon dioxide (CO_2), hydrogen sulfide (H_2S) and mercaptans such as methanethiol (CH_3SH) and ethanethiol (C_2H_5SH).
- Other gases: nitrogen (N_2) and helium (He).
- Water: water vapor and liquid water. Also dissolved salts and dissolved gases (acids).
- Liquid hydrocarbons: perhaps some natural-gas condensate (also referred to as *casinghead gasoline* or *natural gasoline*) and/or crude oil.
- Mercury: very small amounts of mercury primarily in elemental form, but chlorides and other species are possibly present.
- Naturally occurring radioactive material (NORM): natural gas may contain radon, and the produced water may contain dissolved traces of radium, which can accumulate within piping and processing equipment. This can render piping and equipment radioactive over time.

The raw natural gas must be purified to meet the quality standards specified by the major pipeline transmission and distribution companies. Those quality standards vary from pipeline to pipeline and are usually a function of a pipeline system's design and the markets that it serves. In general, the standards specify that the natural gas:

- Be within a specific range of heating value (caloric value). For example, in the United States, it should be about 1035 ± 5% BTU per cubic foot of gas at 1 atmosphere and 60°F (41 MJ ± 5% per cubic metre of gas at 1 atmosphere and 15.6°C).
- Be delivered at or above a specified hydrocarbon dew point temperature (below which some of the hydrocarbons in the gas might condense at pipeline pressure forming liquid slugs that could damage the pipeline).
- Dew-point adjustment serves the reduction of the concentration of water and heavy hydrocarbons in natural gas to such an extent that no condensation occurs during the ensuing transport in the pipelines
- Be free of particulate solids and liquid water to prevent erosion, corrosion or other damage to the pipeline.
- Be dehydrated of water vapor sufficiently to prevent the formation of methane hydrates within the gas processing plant or subsequently within the sales gas transmission pipeline. A typical water content specification in the U.S. is that gas must contain no more than seven pounds of water per million standard cubic feet (MMSCF) of gas.

- Contain no more than trace amounts of components such as hydrogen sulfide, carbon dioxide, mercaptans, and nitrogen. The most common specification for hydrogen sulfide content is 0.25 grain H_2S per 100 cubic feet of gas, or approximately 4 ppm. Specifications for CO_2 typically limit the content to no more than two or three percent.
- Maintain mercury at less than detectable limits (approximately 0.001 ppb by volume) primarily to avoid damaging equipment in the gas processing plant or the pipeline transmission system from mercury amalgamation and embrittlement of aluminum and other metals.

Description of a Natural-Gas Processing Plant

There are a great many ways in which to configure the various unit processes used in the processing of raw natural gas. The block flow diagram below is a generalized, typical configuration for the processing of raw natural gas from non-associated gas wells. It shows how raw natural gas is processed into sales gas pipelined to the end user markets. It also shows how processing of the raw natural gas yields these byproducts:

- Natural-gas condensate
- Sulfur
- Ethane
- Natural-gas liquids (NGL): propane, butanes and C_5+ (which is the commonly used term for pentanes plus higher molecular weight hydrocarbons)

Raw natural gas is commonly collected from a group of adjacent wells and is first processed at that collection point for removal of free liquid water and natural gas condensate. The condensate is usually then transported to an oil refinery and the water is disposed of as wastewater.

The raw gas is then pipelined to a gas processing plant where the initial purification is usually the removal of acid gases (hydrogen sulfide and carbon dioxide). There are many processes that are available for that purpose as shown in the flow diagram, but amine treating is the process that was historically used. However, due to a range of performance and environmental constraints of the amine process, a newer technology based on the use of polymeric membranes to separate the carbon dioxide and hydrogen sulfide from the natural gas stream has gained increasing acceptance. Membranes are attractive since no reagents are consumed.

The acid gases, if present, are removed by membrane or amine treating can then be routed into a sulfur recovery unit which converts the hydrogen sulfide in the acid gas into either elemental sulfur or sulfuric acid. Of the processes available for these conversions, the Claus process is by far the most well known for recovering elemental sulfur, whereas the conventional Contact process and the WSA (Wet sulfuric acid process) are the most used technologies for recovering sulfuric acid.

The residual gas from the Claus process is commonly called *tail gas* and that gas is then processed in a tail gas treating unit (TGTU) to recover and recycle residual sulfur-containing compounds back into the Claus unit. Again, as shown in the flow diagram, there are a number of processes available for treating the Claus unit tail gas and for that purpose a WSA process is also very suitable since it can work autothermally on tail gases.

The next step in the gas processing plant is to remove water vapor from the gas using either the regenerable absorption in liquid triethylene glycol (TEG), commonly referred to as glycol dehydration, deliquescent chloride desiccants, and or a Pressure Swing Adsorption (PSA) unit which is regenerable adsorption using a solid adsorbent. Other newer processes like membranes may also be considered.

Mercury is then removed by using adsorption processes (as shown in the flow diagram) such as activated carbon or regenerable molecular sieves.

Although not common, nitrogen is sometimes removed and rejected using one of the three processes indicated on the flow diagram:

- Cryogenic process (Nitrogen Rejection Unit), using low temperature distillation. This process can be modified to also recover helium, if desired (industrial gas).
- Absorption process, using lean oil or a special solvent as the absorbent.
- Adsorption process, using activated carbon or molecular sieves as the adsorbent. This process may have limited applicability because it is said to incur the loss of butanes and heavier hydrocarbons.

The next step is to recover the natural gas liquids (NGL) for which most large, modern gas processing plants use another cryogenic low temperature distillation process involving expansion of the gas through a turbo-expander followed by distillation in a demethanizing fractionating column. Some gas processing plants use lean oil absorption process rather than the cryogenic turbo-expander process.

The recovered NGL stream is sometimes processed through a fractionation train consisting of three distillation towers in series: a deethanizer, a depropanizer and a debutanizer. The overhead product from the deethanizer is ethane and the bottoms are fed to the depropanizer. The overhead product from the depropanizer is propane and the bottoms are fed to the debutanizer. The overhead product from the debutanizer is a mixture of normal and iso-butane, and the bottoms product is a C_5+ mixture. The recovered streams of propane, butanes and C_5+ may be "sweetened" in a Merox process unit to convert undesirable mercaptans into disulfides and, along with the recovered ethane, are the final NGL by-products from the gas processing plant. Currently, most cryogenic plants do not include fractionation for economic reasons, and the NGL stream is instead transported as a mixed product to standalone fractionation complexes located near refineries or chemical plants that use the components for feedstock. In case laying pipeline is not possible for geographical reason,or the distance between source and consumer exceed 3000 km, natural gas is then transported by ship as LNG (liquefied natural gas) and again converted into its gaseous state in the vicinity of the consumer.

The residue gas from the NGL recovery section is the final, purified sales gas which is pipelined to the end-user markets. Rules and agreements are made between buyer and seller regarding the quality of the gas. These usually specify the maximum allowable concentration of CO_2, H_2S and H_2O as well as requiring the gas to be commercially free from objectionable odours and materials, and dust or other solid or liquid matter, waxes, gums and gum forming constituents, which might damage or adversely affect operation of the buyers equipment. When an upset occurs on the treatment plant buyers can usually refuse to accept the gas, lower the flow rate or re-negotiate the price.

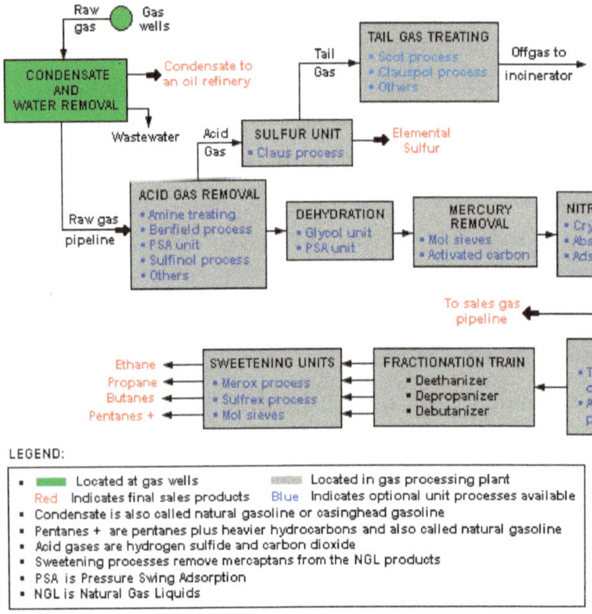

Helium Recovery

If the gas contains significant helium content, the helium may be recovered by fractional distillation. Natural gas may contain as much as 7% helium, and is the commercial source of the noble gas. For instance, the Hugoton Gas Field in Kansas and Oklahoma in the United States contains concentrations of helium from 0.3% to 1.9%, which is separated out as a valuable byproduct.

Consumption

Natural gas consumption patterns, across nations, vary based on access. Countries with large reserves tend to handle the raw-material natural gas more generously, while countries with scarce or lacking resources tend to be more economical. Despite the considerable findings, the predicted availability of the natural-gas reserves has hardly changed.

Applications of Natural Gas

- Fuel for industrial heating and desiccation process
- Fuel for the operation of public and industrial power stations
- Household fuel for cooking, heating and providing hot water
- Fuel for environmentally friendly compressed or liquid natural gas vehicles

- Raw material for chemical synthesis
- Raw material for large-scale fuel production using gas-to-liquid (GTL) process (e.g. to produce sulphur-and aromatic-free diesel with low-emission combustion)

MAGNETIC SEPARATION

Magnetic separation is a process in which magnetically susceptible material is extracted from a mixture using a magnetic force. This separation technique can be useful in mining iron as it is attracted to a magnet. In mines where wolframite was mixed with cassiterite, such as South Crofty and East Pool mine in Cornwall or with bismuth such as at the Shepherd and Murphy mine in Moina, Tasmania, magnetic separation was used to separate the ores. At these mines a device called a Wetherill's Magnetic Separator (invented by John Price Wetherill, 1844–1906) was used. In this machine, the raw ore, after calcination was fed onto a moving belt which passed underneath two pairs of electromagnets under which further belts ran at right angles to the feed belt. The first pair of electromagnets was weakly magnetized and served to draw off any iron ore present. The second pair were strongly magnetized and attracted the wolframite, which is weakly magnetic. These machines were capable of treating 10 tons of ore a day.

It is also used in electromagnetic cranes that separate magnetic material from scraps.

PRECIPITATION (CHEMISTRY)

Precipitation is the creation of a solid from a solution. When the reaction occurs in a liquid solution, the solid formed is called the 'precipitate'. The chemical that causes the solid to form is called the 'precipitant'. Without sufficient force of gravity (settling) to bring the solid particles together, the precipitate remains in suspension. After sedimentation, especially when using a centrifuge to press it into a compact mass, the precipitate may be referred to as a 'pellet'. Precipitation can be used as a medium. The precipitate-free liquid remaining above the solid is called the 'supernate› or ‹supernatant›. Powders derived from precipitation have also historically been known as ‹flowers›. When the solid appears in the form of cellulose fibers which have been through chemical processing, the process is often referred to as **regeneration**.

Sometimes the formation of a precipitate indicates the occurrence of a chemical reaction. If silver nitrate solution is poured into a solution of sodium chloride, a chemical reaction occurs forming a white precipitate of silver chloride. When potassium iodide solution reacts with lead(II) nitrate solution, a yellow precipitate of lead(II) iodide is formed.

Precipitation may occur if the concentration of a compound exceeds its solubility (such as when mixing solvents or changing their temperature). Precipitation may occur rapidly from a supersaturated solution.

In solids, precipitation occurs if the concentration of one solid is above the solubility limit in the host solid, due to e.g. rapid quenching or ion implantation, and the temperature is high enough that diffusion can lead to segregation into precipitates. Precipitation in solids is routinely used to synthesize nanoclusters.

An important stage of the precipitation process is the onset of nucleation. The creation of a hypothetical solid particle includes the formation of an interface, which requires some energy based on the relative surface energy of the solid and the solution. If this energy is not available, and no suitable nucleation surface is available, supersaturation occurs.

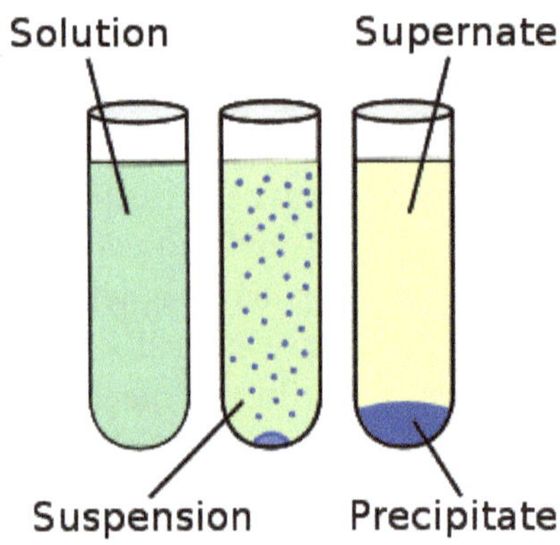

Chemical Precipitation

Applications

Crystals of *meso*-tetratolylporphyrin from a reflux of propanoic acid precipitate on cooling

Copper from a wire is displaced by silver in a silver nitrate solution it is dipped into, and solid silver precipitates out.

Precipitation reactions can be used for making pigments, removing salts from water in water treatment, and in classical qualitative inorganic analysis.

Precipitation is also useful to isolate the products of a reaction during workup. Ideally, the product of the reaction is insoluble in the reaction solvent. Thus, it precipitates as it is formed, preferably forming pure crystals. An example of this would be the synthesis of porphyrins in refluxing propionic acid. By cooling the reaction mixture to room temperature, crystals of the porphyrin precipitate, and are collected by filtration:

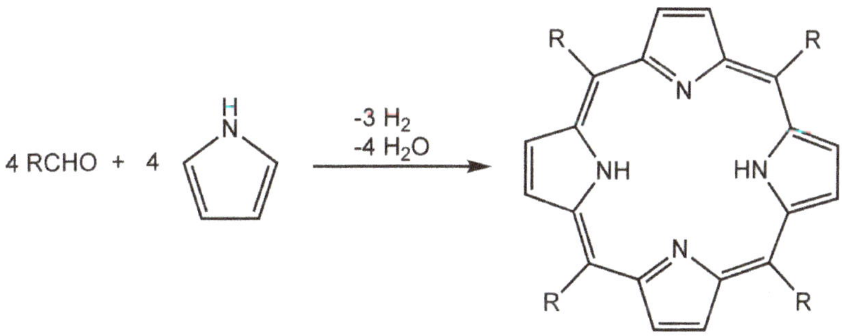

Precipitation may also occur when an *antisolvent* (a solvent in which the product is insoluble) is added, drastically reducing the solubility of the desired product. Thereafter, the precipitate may easily be separated by filtration, decanting, or centrifugation. An example would be the synthesis of chromic tetraphenylporphyrin chloride: water is added to the DMF reaction solution, and the product precipitates.Precipitation is also useful in purifying products: crude bmim-Cl is taken up in acetonitrile, and dropped into ethyl acetate, where it precipitates. Another important application of an antisolvent is in ethanol precipitation of DNA.

In metallurgy, precipitation from a solid solution is also a useful way to strengthen alloys; this process is known as solid solution strengthening.

Representation Using Chemical Equations

An example of a precipitation reaction: Aqueous silver nitrate ($AgNO_3$) is added to a solution containing potassium chloride (KCl), the precipitation of a white solid, silver chloride (AgCl), is observed. (Zumdahl, 2005)

$$AgNO_3 + KCl \longrightarrow AgCl\downarrow + KNO_3$$

The silver chloride (AgCl) has formed a solid, which is observed as a precipitate.

This reaction can be written emphasizing the dissociated ions in a combined solution. This is known as the ionic equation.

$$Ag^+ + NO_3^- + K^+ + Cl^- \longrightarrow AgCl\downarrow + K^+ + NO_3^-$$

A final way to represent a precipitate reaction is known as a *net ionic reaction*. In this case, any spectator ions (those that do not contribute to the reaction) are left out of the formula completely. This simplifies the above equations to the following:

$$Ag^+ + Cl^- \longrightarrow AgCl\downarrow$$

Precipitate Colors

Green and reddish brown stains on a limestone core sample, respectively corresponding to precipitates of oxides/hydroxides of Fe2+ and Fe3+.

Many compounds containing metal ions produce precipitates with distinctive colors. The following are typical colors for various metals. However, many of these compounds can produce colors very different from those listed.

Gold	black
Chromium	Deep green, murky green, orange, purple, yellow, brown
Cobalt	Pink
Copper	Blue
Iron(II)	Green
Iron(III)	Reddish brown
Manganese	Pale pink
Nickel	Green
Lead	Yellow

Other compounds generally form white precipitates.

Anion/Cation Analysis

Precipitate formation is useful in the detection of the type of cation in a salt. To do this, an alkali first reacts with the unknown salt to produce a precipitate that is the hydroxide of the unknown salt. To identify the cation, the color of the precipitate and its solubility in excess are noted. Similar processes are often used in sequence - for example, a barium nitrate solution will react with sulfate ions to form a solid barium sulfate precipitate, indicating that it is likely that sulfate ions are present.

Digestion

Digestion, or *precipitate ageing*, happens when a freshly formed precipitate is left, usually at a higher temperature, in the solution from which it precipitates. It results in cleaner and bigger particles. The physico-chemical process underlying digestion is called Ostwald ripening.

RECRYSTALLIZATION (CHEMISTRY)

In chemistry, recrystallization is a technique used to purify chemicals. By dissolving both impurities and a compound in an appropriate solvent, either the desired compound or impurities can be coaxed out of solution, leaving the other behind. It is named for the crystals often formed when the compound precipitates out. Alternatively, recrystallization can refer to the natural growth of larger ice crystals at the expense of smaller ones.

Chemistry

In chemistry, recrystallization is a procedure for purifying compounds. The most typical situation is that a desired "compound A" is contaminated by a small amount of "impurity B". There are various methods of purification that may be attempted (see Separation process), recrystallization being one of them. There are also different recrystallization techniques that can be used such as:

Single-Solvent Recrystallization

Typically, the mixture of "compound A" and "impurity B" are dissolved in the smallest amount of hot solvent to fully dissolve the mixture, thus making a saturated solution. The solution is then allowed to cool. As the solution cools the solubility of compounds in solution drops. This results in the desired compound dropping (recrystallizing) from solution. The slower the rate of cooling, the bigger the crystals form.

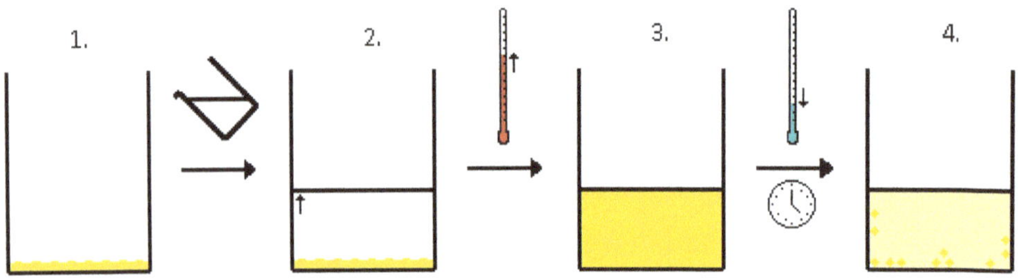

→ Solvent added (clear) to compound (orange) → Solvent heated to give saturated compound solution (orange) → Saturated compound solution (orange) allowed to cool over time to give crystals (orange) and a saturated solution (pale-orange).

Crystallization of Ibuprofen in HCl(aq)

In an ideal situation the solubility product of the impurity, B, is not exceeded at any temperature. In that case the solid crystals will consist of pure A and all the impurity will remain in solution. The solid crystals are collected by filtration and the filtrate is discarded. If the solubility product of the impurity is exceeded, some of the impurity will co-precipitate. However, because of the relatively low concentration of the impurity, its concentration in the precipitated crystals will be less than its concentration in the original solid. Repeated recrystallization will result in an even purer crystalline precipitate. The purity is checked after each recrystallization by measuring the melting point, since impurities lower the melting point. NMR spectroscopy can also be used to check the level of impurity. Repeated recrystallization results in some loss of material because of the non-zero solubility of compound A.

The crystallization process requires an initiation step, such as the addition of a "seed" crystal. In the laboratory a minuscule fragment of glass, produced by scratching the side of the glass recrystallization vessel, may provide the nucleus on which crystals may grow. Successful recrystallization depends on finding the right solvent. This is usually a combination of prediction/experience and trial/error. The compounds must be more soluble at the higher temperature than at the lower temperatures. Any insoluble impurity is removed by the technique of hot filtration.

Multi-Solvent Recrystallization

This method is the same as the above but where two (or more) solvents are used. This relies on both "compound A" and "impurity B" being soluble in a first solvent. A second solvent is slowly added. Either "compound A" or "impurity B" will be insoluble in this solvent and precipitate, whilst the other of "compound A"/"impurity B" will remain in solution. Thus the proportion of first and second solvents is critical. Typically the second solvent is added slowly until one of the compounds begins to crystallize from solution and then the solution is cooled. Heating is not required for this technique but can be used.

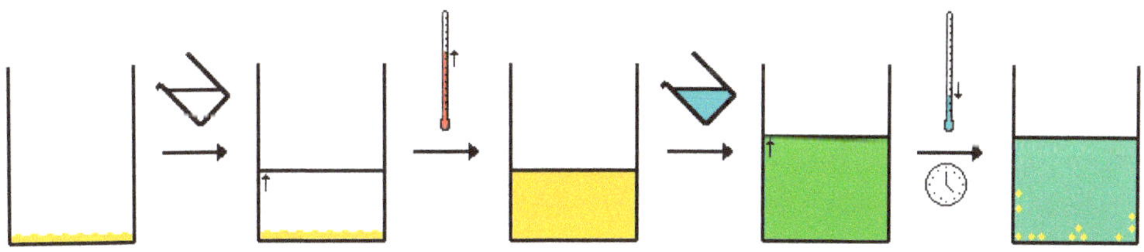

→ Solvent added (clear) to compound (orange) → Solvent heated to give saturated compound solution (orange) → Second solvent (blue) added to compound solution (orange) to give mixed solvent system (green) → Mixed solvent system (green) allowed to cool over time to give crystals (orange) and a saturated mixed solvent system (green-blue).

The reverse of this method can be used where a mixture of solvent dissolves both A and B. One of the solvents is then removed by distillation or by an applied vacuum. This results in a change in the proportions of solvent causing either "compound A" or "impurity B" to precipitate.

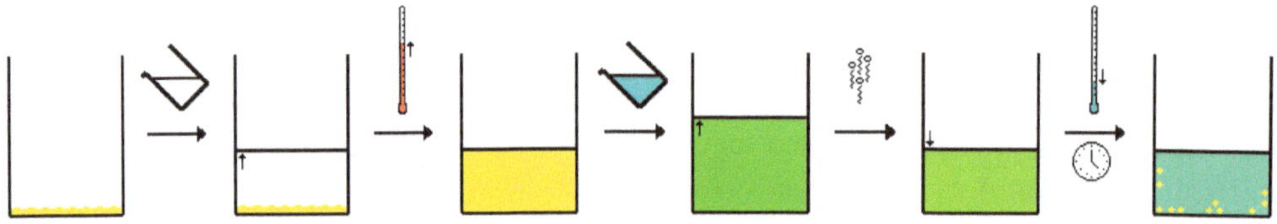

→ First solvent added (clear) to compound (orange) → Solvent heated to give saturated compound solution (orange) → Second solvent (blue) added to compound solution (orange) to give first mixed solvent system (green) → Volatile first solvent (clear) is removed (e.g. evaporation) from first mixed solvent system (green) to give a second mixed solvent system (dark-green) → Second mixed solvent system (dark-green) allowed to cool over time to give crystals (orange) and a saturated second mixed solvent system (green-blue).

Hot Filtration-Recrystallization

Hot filtration can be used to separate "compound A" from both "impurity B" and some "insoluble matter C". This technique normally uses a single-solvent system as described above. When both "compound A" and "impurity B" are dissolved in the minimum amount of hot solvent, the solution is filtered to remove "insoluble matter C". This matter may be anything from a third impurity compound to fragments of broken glass. For a successful procedure, one must ensure that the filtration apparatus is hot in order to stop the dissolved compounds crystallizing from solution during filtration, thus forming crystals on the filter paper or funnel.

One way to achieve this is to heat a conical flask containing a small amount of clean solvent on a hot plate. A filter funnel is rested on the mouth, and hot solvent vapors keep the stem warm. Jacketed filter funnels may also be used. The filter paper is preferably fluted, rather than folded into a quarter; this allows quicker filtration, thus less opportunity for the desired compound to cool and crystallize from the solution.

Often it is simpler to do the filtration and recrystallization as two independent and separate steps. That is dissolve "compound A" and "impurity B" in a suitable solvent at room temperature, filter (to remove insoluble compound/glass), remove the solvent and then recrystallize using any of the methods listed above.

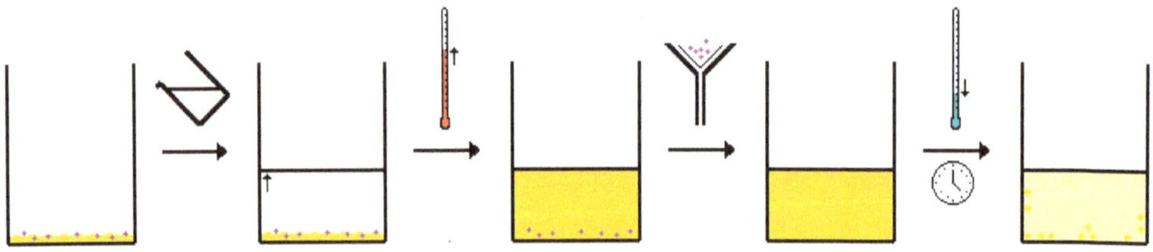

→ Solvent added (clear) to a mixture of compound (orange) + insoluble substance (purple) → Solvent heated to give saturated compound solution (orange) + insoluble substance (purple) → Saturated compound solution (orange) filtered to remove insoluble substance (purple) → Saturated compound solution (orange) allowed to cool over time to give crystals (orange) and a saturated solution (pale-orange).

Seeding

Crystallization requires an initiation step. This can be spontaneous or can be done by adding a small amount of the pure compound (a seed crystal) to the saturated solution, or can be done by simply scratching the glass surface to create a seeding surface for crystal growth. It is thought that even dust particles can act as simple seeds.

Single Perfect Crystals (For X-Ray Analysis)

Growing crystals for X-ray crystallography can be quite difficult. For X-ray analysis, single perfect crystals are required. Typically a small amount (5–100 mg) of pure compound is used, and crystals are allowed to grow very slowly. Several techniques can be used to grow these perfect crystals:

- Slow evaporation of a single solvent - typically the compound is dissolved in a suitable solvent and the solvent is allowed to slowly evaporate. Once the solution is saturated crystals can form.

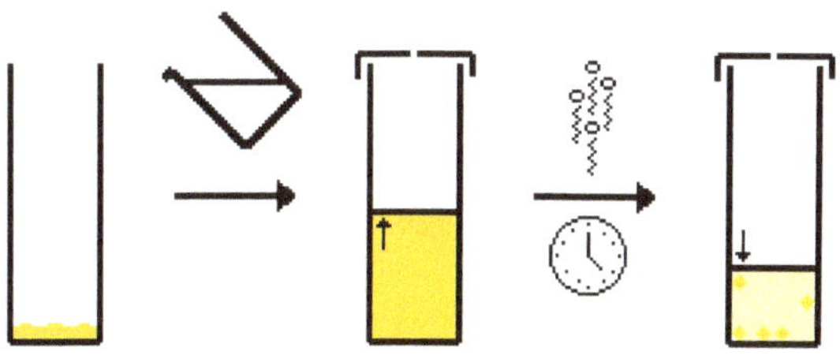

→ Solvent added (clear) to compound (orange) to give compound solution (orange) → Vessel sealed but a small hole allows solvent vapour (clear) to slowly evaporate from compound solution (orange) over time to give crystals (orange) and a saturated solution (pale-orange).

- Slow evaporation of a multi-solvent system - the same as above, however as the solvent composition changes due to evaporation of the more volatile solvent. The compound is more soluble in the volatile solvent, and so the compound becomes increasingly insoluble in solution and crystallizes.

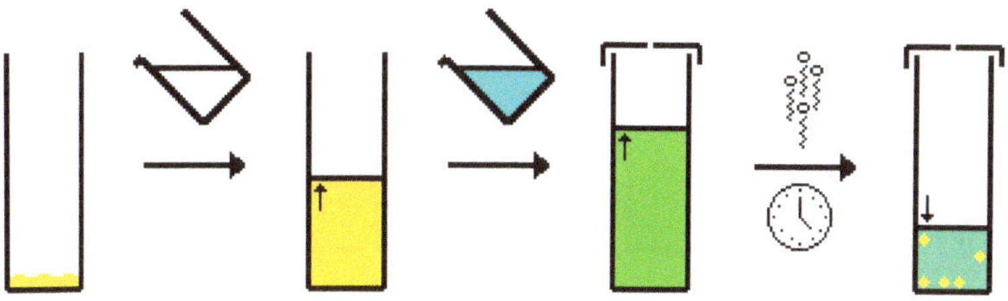

→ Solvent added (clear) to compound (orange) to give compound solution (orange) → Second solvent added (blue) to compound solution (orange) to give mixed solvent system (green) → Vessel sealed but a small hole allows solvent vapour (clear) to slowly evaporate over time to give crystals (orange) and a saturated mixed solvent solution (blue-green).

- Slow diffusion - similar to the above. However, a second solvent is allowed to evaporate from one container into a container holding the compound solution (gas-diffusion). As the solvent composition changes due to an increase in solvent that has gas-diffused into solution, the compound become increasingly insoluble in solution and crystallizes.

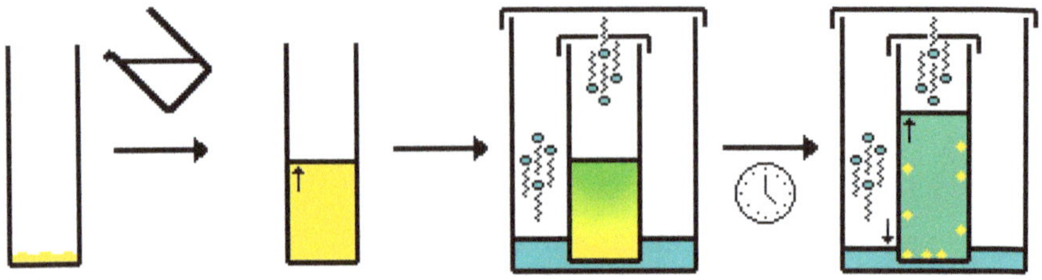

→ Solvent added (clear) to compound (orange) in first vessel to give compound solution (orange) → First vessel is placed in a second vessel contain second solvent (blue). The second vessel is sealed, the first vessel is also sealed, although a small hole in the first vessel is present. This hole allows volatile solvent vapour (blue) to slowly evaporate from second vessel and condensate (that is infuse) into the first vessel, to give a mixed solvent system (green) → Over time this gives crystals (orange) and a saturated mixed solvent system (green-blue).

- Interface/slow mixing (often performed in an NMR tube). Similar to the above, but instead of one solvent gas-diffusing into another, the two solvents mix (diffuse) by liquid-liquid diffusion. Typically a second solvent is "layered" carefully on top of the solution containing the compound. Over time the two solution mix. As the solvent composition changes due to diffusion, the compound becomes increasingly insoluble in solution and crystallizes, usually at the interface. Additionally, it is better to use a denser solvent as the lower layer, and/or a hotter solvent as the upper layer because this results in the slower mixing of the solvents.

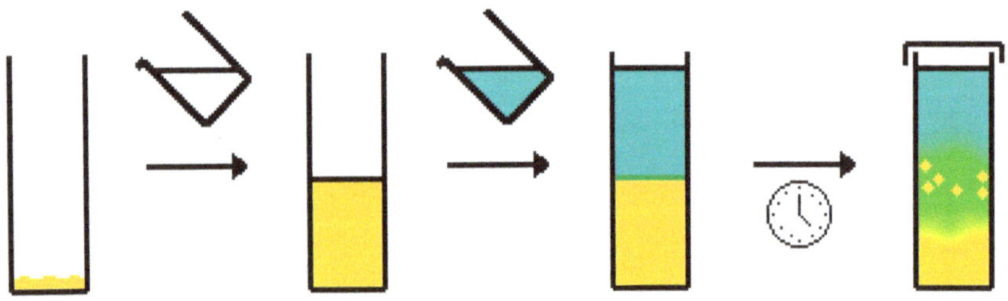

→ Solvent added (clear) to compound (orange) to give compound solution (orange) → Second solvent added (blue) carefully so that the two solvents do not mix. → The two solvents mix (diffuse) slowly over time to give crystals (orange) at solvent interface (green)

- Specialized equipment can be used in the shape of a "H" to perform the above, where one of the vertical line of the "H" is a tube containing a solution of the compound, and the other vertical line of the "H" is a tube containing a solvent which the compound is not soluble in, and the horizontal line of the "H" is a tube which joins the two vertical tubes, which also has a fine glass sinter that restricts the mixing of the two solvents.

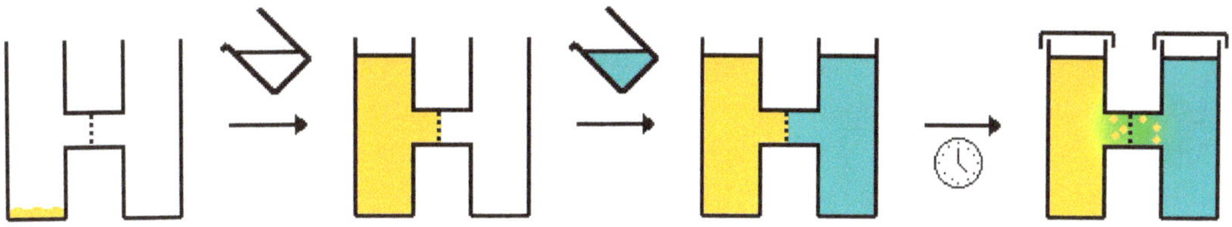

→ Solvent added (clear) to compound (orange) to give a compound solution (orange) → Second solvent added (blue) to the second tube chamber → The two solvents mix slowly over time, the mixing is slowed by a fine sinter separating the two solvent chambers, to give crystals (orange) at solvent interface (green) over time

- Once single perfect crystals have been obtained, it is recommended that the crystals are kept in a sealed vessel with some of the liquid of crystallisation to prevent the crystal from 'drying out'. Single perfect crystals may contain solvent of crystallisation in the crystal lattice. Loss of this internal solvent from the crystals can result in the crystal lattice breaking down, and the crystals turning to powder.

Ice

For ice, recrystallization refers to the growth of larger crystals at the expense of smaller ones. Some biological antifreeze proteins have been shown to inhibit this process, and the effect may be relevant in freezing-tolerant organisms.

SCRUBBER

Scrubber systems are a diverse group of air pollution control devices that can be used to remove some particulates and/or gases from industrial exhaust streams. The first air scrubber was designed to remove carbon dioxide from the air of an early submarine, the Ictineo I, a role for which they continue to be used till today. Traditionally, the term "scrubber" has referred to pollution control devices that use liquid to wash unwanted pollutants from a gas stream. Recently, the term has also been used to describe systems that inject a dry reagent or slurry into a dirty exhaust stream to "wash out" acid gases. Scrubbers are one of the primary devices that control gaseous emissions, especially acid gases. Scrubbers can also be used for heat recovery from hot gases by flue-gas condensation.

There are several methods to remove toxic or corrosive compounds from exhaust gas and neutralize it.

Combustion

Combustion is sometimes the cause of harmful exhausts, but, in many cases, combustion may also be used for exhaust gas cleaning if the temperature is high enough and enough oxygen is available.

Wet Scrubbing

The exhaust gases of combustion may contain substances considered harmful to the environment, and the scrubber may remove or neutralize those. A wet scrubber is used for cleaning air, fuel gas or other gases of various pollutants and dust particles. Wet scrubbing works via the contact of target compounds or particulate matter with the scrubbing solution. Solutions may simply be water (for dust) or solutions of reagents that specifically target certain compounds.

Process exhaust gas can also contain water-soluble toxic and/or corrosive gases like hydrochloric acid (HCl) or ammonia (NH_3). These can be removed very well by a wet scrubber.

Removal efficiency of pollutants is improved by increasing residence time in the scrubber or by the increase of surface area of the scrubber solution by the use of a spray nozzle, packed towers or an aspirator. Wet scrubbers may increase the proportion of water in the gas, resulting in a visible stack plume, if the gas is sent to a stack.

Wet scrubbers can also be used for heat recovery from hot gases by flue-gas condensation. In this mode, termed a condensing scrubber, water from the scrubber drain is circulated through a cooler to the nozzles at the top of the scrubber. The hot gas enters the scrubber at the bottom. If the gas temperature is above the water dew point, it is initially cooled by evaporation of water drops. Further cooling cause water vapors to condense, adding to the amount of circulating water.

The condensation of water release significant amounts of low temperature heat (more than 2 gigajoules (560 kWh) per ton of water), that can be recovered by the cooler for e.g. district heating purposes.

Excess condensed water must continuously be removed from the circulating water.

The gas leaves the scrubber at its dew point, so even though significant amounts of water may have been removed from the cooled gas, it is likely to leave a visible stack plume of water vapor.

Dry Scrubbing

A dry or semi-dry scrubbing system, unlike the wet scrubber, does not saturate the flue gas stream that is being treated with moisture. In some cases no moisture is added, while in others only the amount of moisture that can be evaporated in the flue gas without condensing is added. Therefore, dry scrubbers generally do not have a stack steam plume or wastewater handling/disposal requirements. Dry scrubbing systems are used to remove acid gases (such as SO_2 and HCl) primarily from combustion sources.

There are a number of dry type scrubbing system designs. However, all consist of two main sections or devices: a device to introduce the acid gassorbent material into the gas stream and a particulate matter control device to remove reaction products, excess sorbent material as well as any particulate matter already in the flue gas.

Dry scrubbing systems can be categorized as dry sorbent injectors (DSIs) or as spray dryer absorbers (SDAs). Spray dryer absorbers are also called semi-dry scrubbers or spray dryers.

Dry scrubbing systems are often used for the removal of odorous and corrosive gases from wastewater treatment plant operations. The medium used is typically an activated alumina compound impregnated with materials to handle specific gases such as hydrogen sulfide. Media used can be mixed together to offer a wide range of removal for other odorous compounds such as methyl mercaptans, aldehydes, volatile organic compounds, dimethyl sulfide, and dimethyl disulfide.

Dry sorbent injection involves the addition of an alkaline material (usually hydrated lime, soda ash, or sodium bicarbonate) into the gas stream to react with the acid gases. The sorbent can be injected directly into several different locations: the combustion process, the flue gas duct (ahead of the particulate control device), or an open reaction chamber (if one exists). The acid gases react with the alkaline sorbents to form solid salts which are removed in the particulate control device. These simple systems can achieve only limited acid gas (SO_2 and HCl) removal efficiencies. Higher collection efficiencies can be achieved by increasing the flue gas humidity (i.e., cooling using water spray). These devices have been used on medical waste incinerators and a few municipal waste combustors.

In **spray dryer absorbers**, the flue gases are introduced into an absorbing tower (dryer) where the gases are contacted with a finely atomized alkaline slurry. Acid gases are absorbed by the slurry mixture and react to form solid salts which are removed by the particulate control device. The heat of the flue gas is used to evaporate all the water droplets, leaving a non-saturated flue gas to exit the absorber tower. Spray dryers are capable of achieving high (80+%) acid gas removal efficiencies. These devices have been used on industrial and utility boilers and municipal waste incinerators.

Absorber

Many chemicals can be removed from exhaust gas also by using absorber material. The flue gas is passed through a cartridge which is filled with one or several absorber materials and has been adapted to the chemical properties of the components to be removed. This type of scrubber is sometimes also called dry scrubber. The absorber material has to be replaced after its surface is saturated.

Mercury Removal

Mercury is a highly toxic element commonly found in coal and municipal waste. Wet scrubbers are only effective for removal of soluble mercury species, such as oxidized mercury, Hg^{2+}. Mercury vapor in its elemental form, Hg^0, is insoluble in the scrubber slurry and not removed. Therefore, an additional process of Hg^0 conversion is required to complete mercury capture. Usually halogens are added to the flue gas for this purpose. The type of coal burned as well as the presence of a selective catalytic reduction unit both affect the ratio of elemental to oxidized mercury in the flue gas and thus the degree to which the mercury is removed.

In July 2015, one study found that some mercury scrubbers installed on coal power plants inadvertently capture PAH (polycyclic aromatic hydrocarbons) emissions as well.

Scrubber Waste Products

One side effect of scrubbing is that the process only moves the unwanted substance from the exhaust gases into a liquid solution, solid paste or powder form. This must be disposed of safely, if it can not be reused.

For example, mercury removal results in a waste product that either needs further processing to extract the raw mercury, or must be buried in a special hazardous wastes landfill that prevents the mercury from seeping out into the environment.

As an example of reuse, limestone-based scrubbers in coal-fired power plants can produce a synthetic gypsum of sufficient quality that can be used to manufacture drywall and other industrial products.

Bacteria Spread

Poorly maintained scrubbers have the potential to spread disease-causing bacteria. The problem is a result of inadequate cleaning. For example, the cause of a 2005 outbreak of Legionnaires' disease in Norway was just a few infected scrubbers. The outbreak caused 10 deaths and more than 50 cases of infection.

Sedimentation

Sedimentation is the tendency for particles in suspension to settle out of the fluid in which they are entrained and come to rest against a barrier. This is due to their motion through the fluid in response to the forces acting on them: these forces can be due to gravity, centrifugal acceleration, or electromagnetism. In geology, sedimentation is often used as the opposite of erosion, i.e., the terminal end of sediment transport. In that sense, it includes the termination of transport by saltation or true bedload transport. Settling is the falling of suspended particles through the liquid, whereas sedimentation is the termination of the settling process.

Sedimentation may pertain to objects of various sizes, ranging from large rocks in flowing water to suspensions of dust and pollen particles to cellular suspensions to solutions of single molecules such as proteins and peptides. Even small molecules supply a sufficiently strong force to produce significant sedimentation.

The term is typically used in geology to describe the deposition of sediment which results in the formation of sedimentary rock, but it is also used in various chemical and environmental fields to describe the motion of often-smaller particles and molecules. This process is also used in the biotech industry to separate cells from the culture media.

Experiments

In a sedimentation experiment, the applied force accelerates the particles to a terminal velocity v_{term} at which the applied force is exactly canceled by an opposing drag force. For small enough particles (low Reynolds number), the drag force varies linearly with the terminal velocity, i.e., $F_{drag} = fv_{term}$ (Stokes flow) where f depends only on the properties of the particle and the surrounding fluid. Similarly, the applied force generally varies linearly with some coupling constant (denoted here as q) that depends only on the properties of the particle, $F_{app} = qE_{app}$. Hence, it is generally possible to define a sedimentation coefficient $s \stackrel{\text{def}}{=} q/f$ that depends only on the properties of the particle and the surrounding fluid. Thus, measuring s can reveal underlying properties of the particle.

In many cases, the motion of the particles is blocked by a hard boundary; the resulting accumulation of particles at the boundary is called a sediment. The concentration of particles at the boundary is opposed by the diffusion of the particles.

The sedimentation of a single particle under gravity is described by the Mason–Weaver equation, which has a simple exact solution. The sedimentation coefficient s in this case equals m_b/f, *where* m_b is the buoyant mass.

The sedimentation of a single particle under centrifugal force is described by the Lamm equation, which likewise has an exact solution. The sedimentation coefficient s also equals m_b/f, *where* m_b is the buoyant mass. However, the Lamm equation differs from the Mason–Weaver equation because the centrifugal force depends on radius from the origin of rotation, whereas in the Mason–Weaver equation gravity is constant. The Lamm equation also has extra terms, since it pertains to sector-shaped cells, whereas the Mason–Weaver equation is one-dimensional.

Classification of sedimentation:

- Type 1 sedimentation is characterized by particles that settle discretely at a constant settling velocity,or by the deposition of Iron-Rich minerals to streamlines down to the point source. They settle as individual particles and do not flocculate or stick to other during settling. Example: sand and grit material
- Type 2 sedimentation is characterized by particles that flocculate during sedimentation and because of this their size is constantly changing and therefore their settling velocity is changing. Example: alum or iron coagulation
- Type 3 sedimentation is also known as zone sedimentation. In this process the particles are at a high concentration (greater than 1000 mg/L) such that the particles tend to settle as a mass and a distinct clear zone and sludge zone are present. Zone settling occurs in lime-softening, sedimentation, active sludge sedimentation and sludge thickeners.

Geology

Siltation

In geology, sedimentation is the deposition of particles carried by a fluid flow. For suspended load, this can be expressed mathematically by the Exner equation, and results in the formation of depositional landforms and the rocks that constitute sedimentary record. An undesired increased transport and sedimentation of suspended material is called siltation, and it is a major source of pollution in waterways in some parts of the world. Climate change also affects siltation rates.

Chemistry

In chemistry, sedimentation has been used to measure the size of large molecules (macromolecule), where the force of gravity is augmented with centrifugal force in an ultracentrifuge.

GRAVITY SEPARATION

Gravity separation is an industrial method of separating two components, either a suspension, or dry granular mixture where separating the components with gravity is sufficiently practical: i.e. the components of the mixture have different specific weight. All of the gravitational methods are common in the sense that they all use gravity as the dominant force. The working principle is to lift the material by vacuum over an inclined vibrating screen covered deck. This results in the material being suspended in air while the heavier impurities are left behind on the screen and are discharged from the stone outlet. Gravity separation is used in a wide variety of industries, and can be most simply differentiated by the characteristics of the mixture to be separated - principally that of 'wet' i.e. - a suspension versus 'dry' -a mixture of granular product. Often other methods are applied to make the separation faster and more efficient, such as flocculation, coagulation and suction. The most notable advantages of the gravitational methods are their cost effectiveness and in some cases excellent reduction. Gravity separation is an attractive unit operation as it generally has low capital and operating costs, uses few if any chemicals that might cause environmental concerns and the recent development of new equipment enhances the range of separations possible.

Examples of Application

Agriculture: Gravity Separation tables are used for the removal of impurities, admixture, insect damage and immature kernels from the following examples: Wheat, Barley, Oilseed Rape, Peas, Beans, Cocoa Beans, Linseed. They can be used to separate and standardize coffee beans, cocoa beans, peanuts, corn, peas, rice, wheat, sesame and other food grains.

The Gravity Separator separates products of same size but with difference in specific weight. It has a vibrating rectangular deck, which makes it easy for the product to travel a longer distance, ensuring improved quality of the end product. The pressurized air in the deck enables the material to split according to its specific weight. As a result, the heavier particles travel to the higher level while the lighter particles travel to the lower level of the deck. It comes with easily adjustable air fans to control the volume of air distribution at different areas of the vibrating deck to meet the air supply needs of the deck. The table inclination, speed of eccentric motion and the feed rate can be precisely adjusted to achieve smooth operation of the machine.

Recycling: Gravity Separators are used to remove viable or valuable components from the recycling mixture i.e.: metal from plastic, rubber from plastic, different grades of plastic and these valuable materials are further recycled and reutilized

Preferential Flotation

Heavy liquids such as hhh tetrabromoethane can be used to separate ores from supporting rocks by preferential flotation. The rocks are crushed, and while sand, limestone, dolomite, and other types of rock material will float on TBE, ores such as sphalerite, galena and pyrite will sink.

Clarification/Thickening

Clarification is a name for the method of separating fluid from solid particles. Often clarification is used along with flocculation to make the solid particles sink faster to the bottom of the clarification pool while fluid is obtained from the surface which is free of solid particles.

Thickening is the same as clarification except reverse. Solids that sink to the bottom are obtained and fluid is rejected from the surface.

The difference of these methods could be demonstrated with the methods used in waste water processing: in the clarification phase, sludge sinks to the bottom of the pool and clear water flows over the clear water grooves and continues its journey. The obtained sludge is then pumped into the thickeners, where sludge thickens farther and is then obtained to be pumped into digestion to be prepared into fertilizer.

Sinking Chamber

When clearing gases, an often used and mostly working method for clearing large particles is to blow it into a large chamber where the gas's velocity reduces and the solid particles start sinking to the bottom. This method is used mostly because of its cheap cost.

SIEVE

A sieve, or sifter, is a device for separating wanted elements from unwanted material or for characterizing the particle size distribution of a sample, typically using a woven screen such as a mesh or net or metal. The word "sift" derives from "sieve". In cooking, a sifter is used to separate and break up clumps in dry ingredients such as flour, as well as to aerate and combine them. A strainer is a form of sieve used to separate solids from liquid.

Metal sifters

An ami shakushi, a Japanese ladle or scoop that may be used to remove small drops of batter during the frying of tempura

Industrial Strainer

Some industrial strainers available are simplex basket strainers, duplex basket strainers, and Y strainers. Simple basket strainers are used to protect valuable or sensitive equipment in systems that are meant to be shut down temporarily. Some commonly used strainers are bell mouth strainers, foot valve strainers,basket strainers. Most processing industries (mainly pharmaceutical, coatings and liquid food industries) will opt for a self-cleaning strainer instead of a basket strainer or a simplex strainer due to limitations of simple filtration systems. The self-cleaning strainers or filters are more efficient and provide an automatic filtration solution.

Sieving

Sieving is a simple technique for separating particles of different sizes. A sieve such as used for sifting flour has very small holes. Coarse particles are separated or broken up by grinding against one-another and screen openings. Depending upon the types of particles to be separated, sieves with different types of holes are used. Sieves are also used to separate stones from sand.

Triage sieving refers to grouping people according to their severity of injury.

Wooden Sieves

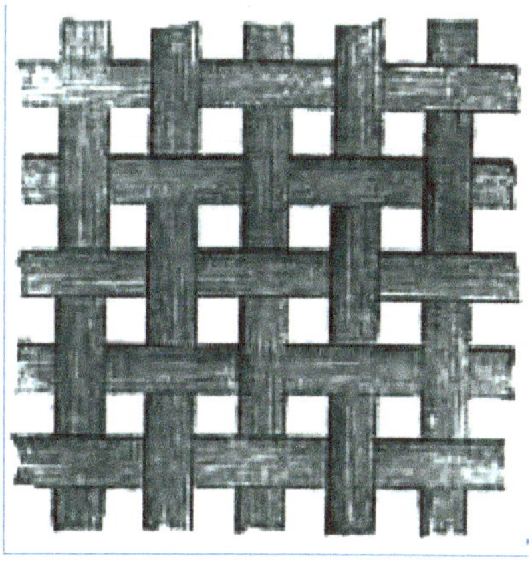

A wooden mesh in which the withes were one eighth of an inch wide and set the same distance apart. This would be used on an English farm of the Victorian era to sift grain, removing dust and soil.

The mesh in a wooden sieve might be made from wood or wicker. Use of wood to avoid contamination is important when the sieve is used for sampling. Henry Stephens, in his *Book of the Farm*, advised that the withes of a wooden riddle or sieve be made from fir or willow with American elm being best. The rims would be made of fir, oak or, especially, beech.

Us Standard Test Sieve Series

A sieve analysis (or gradation test) is a practice or procedure used (commonly used in civil engineering or sedimentology) to assess the particle size distribution (also called gradation) of a granular material. Sieve sizes used in combinations of four to eight sieves. Designations and Nominal Sieve Openings

Tyler (inch/#)	Sieve (inch/#)	Sieve opening (in)	Sieve opcning (mm)
-	5 inch	5.0	125
-	4.24 inch	4.24	106
-	4 inch	4.0	100
-	3-1/2 inch	3.5	90
2.97 inch	3.0 inch	3.0	75
-	2-1/2 inch	2.5	63
-	2.12 inch	2.12	53
2.10 inch	2 inch	2.00	50
-	1-3/4 inch	1.75	45
1.48 inch	1-1/2 inch	1.50	37.5
-	1-1/4 inch	1.25	31.5
1.05 inch	1.06 inch	1.06	26.5
-	1 inch	1.00	25.0
0.883 inch	7/8 inch	0.875	22.4
0.742 inch	3/4 inch	0.750	19.0
0.624 inch	5/8 inch	0.625	16.0
0.525 inch	0.530 inch	0.530	13.2
-	1/2 inch	0.500	12.5
0.441 inch	7/16 inch	0.438	11.2
0.371 inch	3/8 inch	0.375	9.5

Other Types of Sieves

- Chinois, or conical sieve used as a strainer, also sometimes used like a food mill
- Cocktail strainer, a bar accessory
- Colander, a (typically) bowl-shaped sieve used as a strainer in cooking
- Flour sifter or bolter, used in flour production and baking
- Graduated sieves, used to separate varying small sizes of material, often soil, rock or minerals
- Mesh strainer, or just “strainer”, usually consisting of a fine metal mesh screen on a metal frame
- Riddle, used for soil

- Spider, used in Chinese cooking
- Tamis, also known as a drum sieve
- Tea strainer, specifically intended for use when making tea
- *Zaru*, or bamboo sieve, used in Japanese cooking

Stripping (Chemistry)

Stripping is a physical separation process where one or more components are removed from a liquid stream by a vapor steam. In industrial applications the liquid and vapor streams can have co-current or countercurrent flows. Stripping is usually carried out in either a packed or trayed column.

Theory

Stripping works on the basis of mass transfer. The idea is to make the conditions favorable for the component, A, in the liquid phase to transfer to the vapor phase. This involves a gas–liquid interface that A must cross. The total amount of A that has moved across this boundary can be defined as the flux of A, N_A.

Equipment

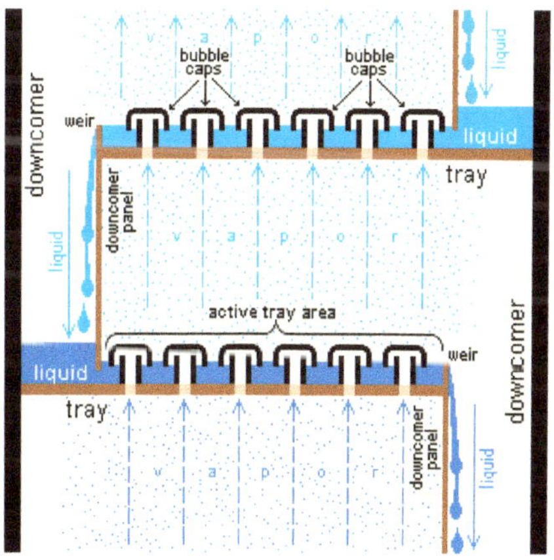

An example of a bubble cap tray that could be found inside of a stripping column.

Stripping is mainly conducted in trayed towers (plate columns) and packed columns, and less often in spray towers, bubble columns, and centrifugal contactors.

Trayed towers consist of a vertical column with liquid flowing in the top and out the bottom. The vapor phase enters in the bottom of the column and exits out of the top. Inside of the column are trays or plates. These trays force the liquid to flow back and forth horizontally while the vapor bubbles up through holes in the trays. The purpose of these trays is to increase the amount of contact area between the liquid and vapor phases.

Packed columns are similar to trayed columns in that the liquid and vapor flows enter and exit in the same manner. The difference is that in packed towers there are no trays. Instead, packing is used to increase the contact area between the liquid and vapor phases. There are many different types of packing used and each one has advantages and disadvantages.

Variables

The variables and design considerations for strippers are many. Among them are the entering conditions, the degree of recovery of the solute needed, the choice of the stripping agent and its flow, the operating conditions, the number of stages, the heat effects, and the type and size of the equipment.

The degree of recovery is often determined by environmental regulations, such as for volatile organic compounds like chloroform.

Frequently, steam, air, inert gases, and hydrocarbon gases are used as stripping agents. This is based on solubility, stability, degree of corrosiveness, and availability. As stripping agents are gases, operation at nearly the highest temperature and lowest pressure that will maintain the components and not vaporize the liquid feed stream is desired. This allows for the minimization of flow. As with all other variables, minimizing cost while achieving efficient separation is the ultimate goal.

The size of the equipment, and particularly the height and diameter, is important in determining the possibility of flow channeling that would reduce the contact area between the liquid and vapor streams. If flow channeling is suspected to be occurring, a redistribution plate is often necessary to, as the name indicates, redistribute the liquid flow evenly to reestablish a higher contact area.

As mentioned previously, strippers can be trayed or packed. Packed columns, and particularly when random packing is used, are usually favored for smaller columns with a diameter less than 2 feet and a packed height of not more than 20 feet. Packed columns can also be advantageous for corrosive fluids, high foaming fluids, when fluid velocity is high, and when particularly low pressure drop is desired. Trayed strippers are advantageous because of ease of design and scale up. Structured packing can be used similar to trays despite possibly being the same material as dumped (random) packing. Using structured packing is a common method to increase the capacity for separation or to replace damaged trays.

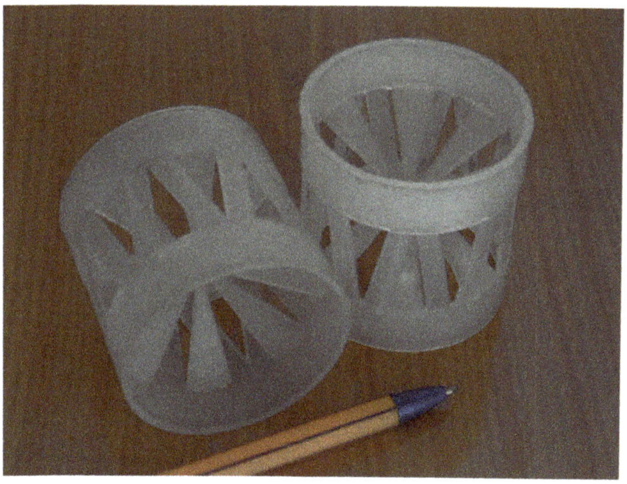

Plastic Białecki rings

Trayed strippers can have sieve, valve, or bubble cap trays while packed strippers can have either structured packing or random packing. Trays and packing are used to increase the contact area over which mass transfer can occur as mass transfer theory dictates. Packing can have varying material, surface area, flow area, and associated pressure drop. Older generation packing include ceramic Raschig rings and Berl saddles. More common packing materials are metal and plastic Pall rings, metal and plastic Zbigniew Białecki rings, and ceramic Intalox saddles. Each packing material of this newer generation improves the surface area, the flow area, and/or the associated pressure drop across the packing. Also important, is the ability of the packing material to not stack on top of itself. If such stacking occurs, it drastically reduces the surface area of the material. Lattice design work has been increasing of late that will further improve these characteristics.

During operation, monitoring the pressure drop across the column can help to determine the performance of the stripper. A changed pressure drop over a significant range of time can be an indication that the packing may need to be replaced or cleaned.

Typical Applications

Stripping is commonly used in industrial applications to remove harmful contaminants from waste streams. One example would be the removal of TBT and PAH contaminants from harbor soils. The soils are dredged from the bottom of contaminated harbors, mixed with water to make a slurry and then stripped with steam. The cleaned soil and contaminant rich steam mixture are then separated. This process is able to decontaminate soils almost completely.

Steam is also frequently used as a stripping agent for water treatment. Volatile organic compounds are partially soluble in water and because of environmental considerations and regulations, must be removed from groundwater, surface water, and wastewater. These compounds can be present because of industrial, agricultural, and commercial activity.

Sublimation (Phase Transition)

Sublimation is the transition of a substance directly from the solid to the gas phase without passing through the intermediate liquid phase. Sublimation is an endothermic phase transition that occurs at temperatures and pressures below a substance's triple point in its phase diagram. The reverse process of sublimation is deposition or desublimation, in which a substance passes directly from a gas to a solid phase. Sublimation has also been used as a generic term to describe a solid-to-gas transition (sublimation) followed by a gas-to-solid transition (deposition).

At normal pressures, most chemical compounds and elements possess three different states at different temperatures. In these cases, the transition from the solid to the gaseous state requires an intermediate liquid state. The pressure referred to is the partial pressure of the substance, not the total (e.g. atmospheric) pressure of the entire system. So, all solids that possess an appreciable vapor pressure at a certain temperature usually can sublimate in air (e.g. water ice just below 0 °C). For some substances, such as carbon and arsenic, sublimation is much easier than evaporation from the melt, because the pressure of their triple point is very high, and it is difficult to obtain them as liquids.

The term sublimation refers to a physical change of state and is not used to describe transformation of a solid to a gas in a chemical reaction. For example, the dissociation on heating of solid ammonium chloride into hydrogen chloride and ammonia is not sublimation but a chemical reaction. Similarly the combustion of candles, containing paraffin wax, to carbon dioxide and water vapor is not sublimation but a chemical reaction with oxygen.

Sublimation requires additional energy and is an endothermic change. The enthalpy of sublimation (also called heat of sublimation) can be calculated by adding the enthalpy of fusion and the enthalpy of vaporization.

Dark green crystals of nickelocene, sublimed from the round-bottom flask and freshly deposited on a cold finger

Examples

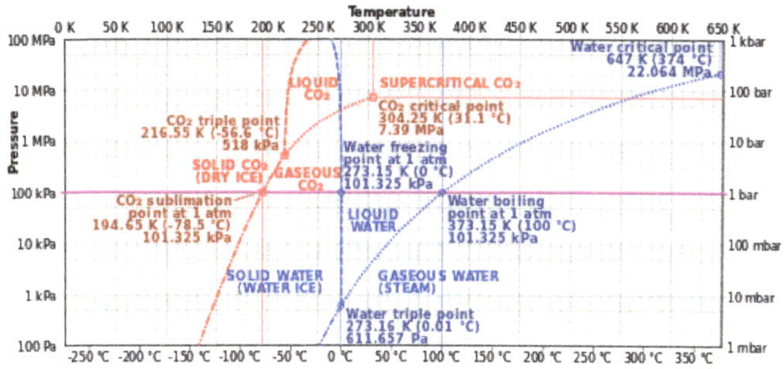

Comparison of phase diagrams of carbon dioxide (red) and water (blue) showing the carbon dioxide sublimation point (middle-left) at 1 atmosphere. As dry ice is heated, it crosses this point along the bold horizontal line from the solid phase directly into the gaseous phase. Water, on the other hand, passes through a liquid phase.

Carbon Dioxide

Small pellets of dry ice subliming in air

Solid carbon dioxide (dry ice) sublimes everywhere along the line below the triple point (e.g., at the temperature of −78.5 °C (194.65 K, −104.2 °F) at atmospheric pressure), whereas its melting into liquid CO_2 can occur only along the line at pressures and temperatures above the triple point (i.e., 5.2 atm, −56.4 °C).

Water

Snow and ice sublime, although more slowly, at temperatures below the freezing/melting point temperature line at 0 °C for most pressures; see line below triple point. In freeze-drying, the material to be dehydrated is frozen and its water is allowed to sublime under reduced pressure or vacuum. The loss of snow from a snowfield during a cold spell is often caused by sunshine acting directly on the upper layers of the snow. Ablation is a process that includes sublimation and erosive wear of glacier ice.

Naphthalene

Naphthalene, an organic compound commonly found in pesticide such as mothball also sublimes. It sublimes easily because it is made of non-polar molecules and has van der Waals intermolecular forces. Naphthalene is a solid that sublimes at standard atmospheric temperature with the sublimation point at around 80°C or 176°F. At low temperature, its vapour pressure is high enough, 1 mmHg at 53°C, to make the solid form of naphthalene evaporate into gas. On the cool surface, the sublimated vapour will be solidified to form a needle-like crystal.

Experimental set up for the sublimation reaction of naphthalene Solid naphthalene sublimes and form the crystal-like structure at the bottom of the watch glass

Solid compound of naphthalene sublimed to form a crystal-like structure on the cool surface.

Other Substances

Camphor subliming in a cold finger. The crude product in the bottom is dark brown; the white purified product on the bottom of the cold finger above is hard to see against the light background.

Iodine produces fumes on gentle heating. It is possible to obtain liquid iodine at atmospheric pressure by controlling the temperature at just above the melting point of iodine. In forensic science, iodine vapor can reveal latent fingerprints on paper. Arsenic can also sublime at high temperatures.

Sublimation Purification

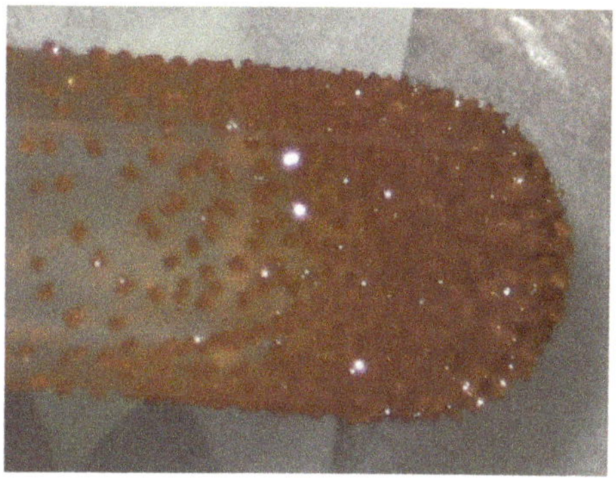

Crystals of ferrocene after purification by vacuum sublimation

Sublimation is a technique used by chemists to purify compounds. A solid is typically placed in a sublimation apparatus and heated under vacuum. Under this reduced pressure, the solid volatilizes and condenses as a purified compound on a cooled surface (cold finger), leaving a non-volatile residue of impurities behind. Once heating ceases and the vacuum is removed, the purified compound may be collected from the cooling surface. For even higher purification efficiencies a temperature gradient is applied, which also allows for the separation of different fractions. Typical setups use an evacuated glass tube that is gradually heated in a controlled manner. The material flow is from the hot end, where the initial material is placed, to the cold end that is connected to a pump stand. By controlling temperatures along the length of the tube the operator can control the zones of recondensation, with very volatile compounds being pumped out of the system completely (or caught by a separate cold trap), moderately volatile compounds recondensating along the tube according to their different volatilities, and non-volatile compounds remaining in the hot end. Vacuum sublimation of this type is also the method of choice for purification of organic compounds for the use in the organic electronics industry, where very high purities (often > 99.99%) are needed to satisfy the standards for consumer electronics and other applications.

In ancient alchemy, a protoscience that contributed to the development of modern chemistry and medicine, alchemists developed a structure of basic laboratory techniques, theory, terminology, and experimental methods. *Sublimation* was used to refer to the process in which a substance is heated to a vapor, then immediately collects as sediment on the upper portion and neck of the heating medium (typically a retort or alembic), but can also be used to describe other similar non-laboratory transitions. It is mentioned by alchemical authors such as Basil Valentine and George Ripley, and in the Rosarium philosophorum, as a process necessary for the completion of the magnum opus. Here, the word *sublimation* is used to describe an exchange of "bodies" and "spirits" similar to laboratory phase transition between solids and gases. Valentine, in his *Triumphal Chariot of Antimony* (published 1678) makes a comparison to spagyrics in which a vegetable sublimation can be used to separate the spirits in wine and beer. Ripley uses language more indicative of the mystical implications of sublimation, indicating that the process has a double aspect in the spiritualization of the body and the corporalizing of the spirit. He writes:

> And Sublimations we make for three causes,The first cause is to make the body spiritual. The second is that the spirit may be corporeal, And become fixed with it and consubstantial. The third cause is that from its filthy original. It may be cleansed, and its saltiness sulphurious May be diminished in it, which is infectious.

Sublimation Predictions

The enthalpy of sublimation has commonly been predicted using the equipartition theorem. If the lattice energy is assumed to be approximately half the packing energy, then the following thermodynamic corrections can be applied to predict the enthalpy of sublimation. Assuming a 1 molar ideal gas gives a correction for the thermodynamic environment (pressure and volume) in which $pV = RT$, hence a correction of 1RT. Additional corrections for the vibrations, rotations and translation then need to be applied. From the equipartition theorem gaseous rotation and translation contribute 1.5RT each to the final state, therefore a +3RT correction. Crystalline vibrations and rotations contribute 3RT each to the initial state, hence −6RT. Summing the RT corrections ; −6RT + 3RT + RT = −2RT. This leads to the following approximate sublimation enthalpy. A similar approximation can be found for the entropy term if rigid bodies are assumed.

Souders–Brown Equation

The Souders–Brown equation (named after Mott Souders and George Granger Brown) has been a tool for obtaining the maximum allowable vapor velocity in vapor–liquid separation vessels (variously called flash drums, knockout drums, knockout pots, compressor suction drums and compressor inlet drums). It has also been used for the same purpose in designing trayed fractionating columns, trayed absorption columns and other vapor–liquid-contacting columns.

A vapor–liquid separator drum is a vertical vessel into which a liquid and vapor mixture (or a flashing liquid) is fed and wherein the liquid is separated by gravity, falls to the bottom of the vessel, and is withdrawn. The vapor travels upward at a design velocity which minimizes the entrainment of any liquid droplets in the vapor as it exits the top of the vessel.

Use

The diameter of a vapor–liquid separator drum is dictated by the expected volumetric flow rate of vapor and liquid from the drum. The following sizing methodology is based on the assumption that those flow rates are known.

Use a vertical pressure vessel with a length–diameter ratio of about 3 to 4, and size the vessel to provide about 5 minutes of liquid inventory between the normal liquid level and the bottom of the vessel (with the normal liquid level being somewhat below the feed inlet).

Calculate the maximum allowable vapor velocity in the vessel by using the Souders–Brown equation:

$$V = (k)\sqrt{\frac{\rho_L - \rho_V}{\rho_V}}$$

where:

V = maximum allowable vapor velocity, m/s

ρ_L = liquid density, kg/m^3

ρ_V = vapor density, kg/m^3

k = 0.107 m/s (when the drum includes a de-entraining mesh pad)

Then the cross-sectional area of the drum (***A***) is obtained from:

A, in m^2 = (vapor flow rate, in m^3/s) ÷ (vapor velocity V, in m/s)

And the drum diameter (***D***) is:

D, in m = [(4) (A) ÷ (3.1416)] $^{0.5}$

The drum should have a vapor outlet at the top, liquid outlet at the bottom, and feed inlet at about the half-full level. At the vapor outlet, provide a de-entraining mesh pad within the drum such that the vapor must pass through that mesh before it can leave the drum. Depending upon how much liquid flow is expected, the liquid outlet line should probably have a liquid level control valve.

As for the mechanical design of the drum (materials of construction, wall thickness, corrosion allowance, etc.), use the same criteria as for any pressure vessel.

Recommended Values of *k*

The *GPSA Engineering Data Book* recommends the following ***k*** values for vertical drums with horizontal mesh pads (at the denoted operating pressures):

- At a gauge pressure of 0 bar: 0.107 m/s
- At a gauge pressure of 7 bar: 0.107 m/s
- At a gauge pressure of 21 bar: 0.101 m/s
- At a gauge pressure of 42 bar: 0.092 m/s
- At a gauge pressure of 63 bar: 0.083 m/s
- At a gauge pressure of 105 bar: 0.065 m/s

GPSA notes:

- ***k*** = 0.107 at a gauge pressure of 7 bar. Subtract 0.003 for every 7 bar above a gauge pressure of 7 bar.
- For glycol or amine solutions, multiply above ***k*** values by 0.6 – 0.8
- Typically use one-half of the above ***k*** values for approximate sizing of vertical separators without mesh pads
- For compressor suction scrubbers and expander inlet separators, multiply ***k*** by 0.7 – 0.8

WINNOWING

Wind winnowing is an agricultural method developed by ancient cultures for separating grain from chaff. It is also used to remove weevils or other pests from stored grain. Threshing, the loosening of grain or seeds from the husks and straw, is the step in the chaff-removal process that comes before winnowing.

In its simplest form it involves throwing the mixture into the air so that the wind blows away the lighter chaff, while the heavier grains fall back down for recovery. Techniques included using a winnowing fan (a shaped basket shaken to raise the chaff) or using a tool (a winnowing fork or shovel) on a pile of harvested grain.

Rice winnowing, Uttarakhand, India

Winnowing in a Dalit village in Tamil Nadu, India

Use of winnowing forks

In Greek culture

The winnowing-fan featured in the rites accorded Dionysus and in the Eleusinian Mysteries: "it was a simple agricultural implement taken over and mysticised by the religion of Dionysus," Jane Ellen Harrison remarked. *Dionysus Liknites* ("Dionysus of the winnowing fan") was wakened by the Dionysian women, in this instance called *Thyiades*, in a cave on Parnassus high above Delphi; the winnowing-fan links the god connected with the mystery religions to the agricultural cycle, but mortal Greek babies too were laid in a winnowing-fan. In Callimachus' *Hymn to Zeus*, Adrasteia lays the infant Zeus in a golden *líknon*, her goat suckles him and he is given honey.

In the Odyssey, the dead oracle Teiresias tells Odysseus to walk away from Ithaca with an oar until a wayfarer tells him it is a winnowing fan (i.e., until Odysseus has come so far from the sea that people don't recognize oars), and there to build a shrine to Poseidon.

In China

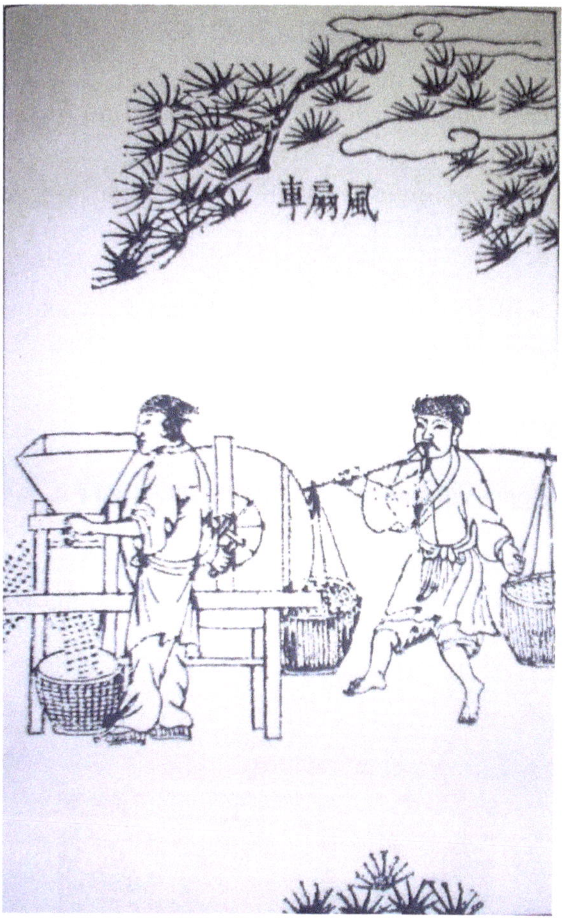

Chinese rotary fan winnowing machine, from the *Tiangong Kaiwu* encyclopedia (1637)

In Ancient China the method was improved by mechanisation with the development of the rotary winnowing fan, which used a cranked fan to produce the airstream. This was featured in Wang Zhen›s book the *Nong Shu* of 1313 AD.

In the Old Testament

In the Old Testament the word winnow is used in several verses in different books in the New International Version while other versions of the bible translate the action as “fan”, “throw” or the separating tool as “pitchfork”, “shovel”, “winnowing fan”, or “winnowing instrument”.

Ruth 3:2 “Now Boaz, with whose women you have worked, is a relative of ours. Tonight he will be winnowing barley on the threshing floor.”

Proverbs 20:8 “When a king sits on his throne to judge, he winnows out all evil with his eyes.”

Proverbs 20:26 "A wise king winnows out the wicked; he drives the threshing wheel over them."

Isaiah 41:16 "You will winnow them, the wind will pick them up, and a gale will blow them away. But you will rejoice in the Lord and glory in the Holy One of Israel."

Jeremiah 4:11 "At that time this people and Jerusalem will be told, "A scorching wind from the barren heights in the desert blows toward my people, but not to winnow or cleanse;"

Jeremiah 15:7 "I will winnow them with a winnowing fork at the city gates of the land. I will bring bereavement and destruction on my people, for they have not changed their ways."

Jeremiah 51:2 "I will send foreigners to Babylon to winnow her and to devastate her land; they will oppose her on every side in the day of her disaster."

In the New Testament

In Matthew 3:12, a sentence introduces the separation of wheat and chaff (good and bad) by "His winnowing fan is in his hand" (American Standard Bible and New American Bible translation). The New International Version, the New Revised Standard Version and the New American Standard Bible translate the term as "winnowing fork"; the Holman Christian Standard Bible translates it as "winnowing shovel".

In Europe

Le vanneur (*The Winnower*) by Jean-François Millet, a 19th-century depiction of winnowing by fan

In Saxon settlements such as one identified in Northumberland as Bede's Ad Gefrin (now called Yeavering) the buildings were shown by an excavator›s reconstruction to have opposed entries. In barns a draught created by the use of these opposed doorways was used in winnowing.

The technique developed by the Chinese was not adopted in Europe until the 18th century, when winnowing machines used a ‹sail fan›. The rotary winnowing fan was exported to Europe, brought there by Dutch sailors between 1700 and 1720. Apparently they had obtained them from the Dutch settlement of Batavia in Java, Dutch East Indies. The Swedes imported some from south China at about the same time and Jesuits had taken several to France from China by 1720. Until the beginning of the 18th century, no rotary winnowing fans existed in the West.

In the United States

The development of the winnowing barn allowed rice plantations in South Carolina to increase their yields dramatically.

Mechanization of the Process

Winnowing machine from 1839

In 1737 Andrew Rodger, a farmer on the estate of Cavers in Roxburghshire, developed a winnowing machine for corn, called a 'Fanner'. These were successful and the family sold them throughout Scotland for many years. Some Scottish Presbyterian ministers saw the fanners as sins against God, for wind was a thing specially made by him and an artificial wind was a daring and impious attempt to usurp what belonged to God alone. As the Industrial Revolution, the winnowing process was mechanized by the invention of additional winnowing machines, such as fanning mills.

ZONE MELTING

Zone melting (or zone refining or floating zone process or travelling melting zone) is a group of similar methods of purifying crystals, in which a narrow region of a crystal is melted, and this molten zone is moved along the crystal. The molten region melts impure solid at its forward edge and leaves a wake of purer material solidified behind it as it moves through the ingot. The impurities concentrate in the melt, and are moved to one end of the ingot. Zone refining was invented by John Desmond Bernal and further developed by William Gardner Pfann in Bell Labs as a method to prepare high purity materials, mainly semiconductors, for manufacturing transistors. Its early use was on germanium for this purpose, but it can be extended to virtually any solute-solvent system having an appreciable concentration difference between solid and liquid phases at equilibrium. This process is also known as the float zone process, particularly in semiconductor materials processing.

(left) Pfann, at left, showing the first zone refining tube, Bell Labs, 1953
(right) Vertical zone refining, 1961. The induction heating coil carrying a high frequency alternating current, melts a section of the metal bar in the tube. The coil moves slowly down the tube, moving the molten zone to the end of the bar.

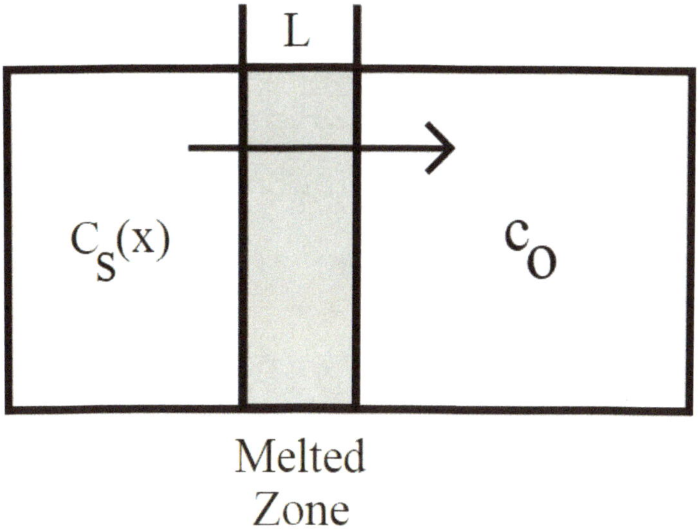

Liquid moves from left to right during melting in the float-zone crystal growth process

Silicon crystal in the beginning of the growth process

Growing silicon crystal

A high-purity (99.999% = 5N) tantalum single crystal, made by the floating zone process (cylindrical object in the center)

Process Details

The principle is that the segregation coefficient *k* (the ratio of an impurity in the solid phase to that in the liquid phase) is usually less than one. Therefore, at the solid/liquid boundary, the impurity atoms will diffuse to the liquid region. Thus, by passing a crystal boule through a thin section of furnace very slowly, such that only a small region of the boule is molten at any time, the impurities will be segregated at the end of the crystal. Because of the lack of impurities in the leftover regions which solidify, the boule can grow as a perfect single crystal if a seed crystal is placed at the base to initiate a chosen direction of crystal growth. When high purity is required, such as in semiconductor industry, the impure end of the boule is cut off, and the refining is repeated.

In zone refining, solutes are segregated at one end of the ingot in order to purify the remainder, or to concentrate the impurities. In **zone leveling**, the objective is to distribute solute evenly throughout the purified material, which may be sought in the form of a single crystal. For example, in the preparation of a transistor or diodesemiconductor, an ingot of germanium is first purified by zone refining. Then a small amount of antimony is placed in the molten zone, which is passed through the pure germanium. With the proper choice of rate of heating and other variables, the antimony can be spread evenly through the germanium. This technique is also used for the preparation of silicon for use in computer chips.

Heaters

A variety of heaters can be used for zone melting, with their most important characteristic being the ability to form short molten zones that move slowly and uniformly through the ingot. Induction coils, ring-wound resistance heaters, or gas flames are common methods. Another method is to pass an electric current directly through the ingot while it is in a magnetic field, with the resulting magnetomotive force carefully set to be just equal to the weight in order to hold the liquid suspended. Optical heaters using high powered halogen or xenon lamps are used extensively in research facilities particularly for the production of insulators, but their use in industry is limited by the relatively low power of the lamps, which limits the size of crystals produced by this method. Zone melting can be done as a batch process, or it can be done continuously, with fresh impure material being continually added at one end and purer material being removed from the other, with impure zone melt being removed at whatever rate is dictated by the impurity of the feed stock.

Indirect-heating floating zone methods use an induction-heated tungsten ring to heat the ingot radiatively, and are useful when the ingot is of a high-resistivity semiconductor on which classical induction heating is ineffective.

Mathematical Expression of Impurity Concentration

When the liquid zone moves by a distance d_x, the number of impurities in the liquid change. Impurities are incorporated in the melting liquid and freezing solid.

k_O: Segregation coefficient

L: Zone length

C_O: Initial uniform impurity concentration of the rod

C_L: Concentration of impurities in the liquid

I: Number of impurities in the liquid

I_O: Number of impurities in zone when first formed at bottom

The number of impurities in the liquid changes in accordance with the express ion below during the movement d_x of the molten zone

$$dI = (C_O - k_O C_L)\, dx$$

$$C_L = I/L$$

$$\int_0^x dx = \int_{I_O}^{I} \frac{dI}{C_O - \frac{k_O I}{L}}$$

$$I_O = C_O L$$

$$C_S = k_O I/L$$

$$C_S(x) = C_O \left(1 - (1 - k_O) e^{-\frac{k_O x}{L}}\right)$$

Applications

Solar Cells

In solar cells float zone processing is particularly useful because the single crystal silicon grown has desirable properties. The bulk charge carrier lifetime in float-zone silicon is the highest among various manufacturing processes. Float-zone carrier lifetimes are around 1000 microseconds compared to 20-200 microseconds with Czochralski process, and 1–30 microseconds with cast multi-crystalline silicon. A longer bulk lifetime increases the efficiency of solar cells significantly.

Related Processes

Zone Remelting

Another related process is **zone remelting**, in which two solutes are distributed through a pure metal. This is important in the manufacture of semiconductors, where two solutes of opposite conductivity type are used. For example, in germanium, pentavalent elements of group V such as antimony and arsenic produce negative (n-type) conduction and the trivalent elements of group III such as aluminum and boron produce positive (p-type) conduction. By melting a portion of such an ingot and slowly refreezing it, solutes in the molten region become distributed to form the desired n-p and p-n junctions.

REFERENCES

1. M. Souders and G. G. Brown (1934). "Design of Fractionating Columns, Entrainment and Capacity". Industrial & Engineering Chemistry. 38 (1): 98–103. doi:10.1021/ie50289a025.
2. Analytical Study of Liquid/Vapour Separation Efficiency, study developed by W.D. Monnery, Chem-Pet Process Technology Ltd. and W.Y. Svrcek, University of Calgary, Calgary, Canada, 2005, for Petroleum Technology Alliance Canada
3. Gas Processing Suppliers Association (GPSA) (1987). Engineering Data Book. 1 (10th ed.). Gas Processing Suppliers Association, Tulsa, Oklahoma.
4. Harrison, Prolegomean to the Study of Greek Religion, 3rd ed. (1922:159).
5. Karl Kerenyi, Dionysus: Archetypal Image of Indestructible Life (1976:44).
6. The Question of the Transmission of the Rotary Winnowing Fan from China to Europe: Some New Findings, Hans Ulrich Vogel, 8th International Conference on the History of Science in China
7. Münzenberg, Hessen. Chapel and Palas (G.Binding, Burg Münzenberg, 1962)
8. M.W.Thompson, The Rise of the Castle, (Cambridge University Press, 1991), 5–6.
9. Broadcasting and winnowing, Antique Farm Tools
10. Robert Temple, The Genius of China, p. 24
11. Chambers, Robert (1885). Domestic Annals of Scotland. Edinburgh : W & R Chambers. p. 397.
12. "Historical Markers - Samuel Wetherill". ExplorePAhistory.com. Retrieved 2012-08-20.
13. Earl, Bryan (1994). Cornish Mining: The Techniques of Metal Mining in the West of England, Past, and Present (2nd ed.). St Austell: Cornish Hillside Publications. p. 97. ISBN 0-9519419-3-3.
14. Dhara, S. (2007). "Formation, Dynamics, and Characterization of Nanostructures by Ion Beam Irradiation". Critical Reviews in Solid State and Materials Sciences. 32 (1): 1–50. doi:10.1080/10408430601187624.
15. A. D. Adler; F. R. Longo; J. D. Finarelli; J. Goldmacher; J. Assour; L. Korsakoff (1967). "A simplified synthesis for meso-tetraphenylporphine". J. Org. Chem. 32 (2): 476–476. doi:10.1021/jo01288a053.
16. Alan D. Adler; Frederick R. Longo; Frank Kampas; Jean Kim (1970). "On the preparation of metalloporphyrins". Journal of Inorganic and Nuclear Chemistry. 32 (7): 2443. doi:10.1016/0022-1902(70)80535-8.
17. Dupont, J., Consorti, C., Suarez, P., de Souza, R. (2004). "Preparation of 1-Butyl-3-methyl imidazolium-based Room Temperature Ionic Liquids". Org. Synth. ; Coll. Vol., 10, p. 184
18. Laurence M. Harwood, Christopher J. Moody (1989). Experimental organic chemistry: Principles and Practice. Oxford: Blackwell Scientific Publications. pp. 127–132. ISBN 0-632-02017-2.

19. Laurence M. Harwood, Christopher J. Moody (1989). Experimental organic chemistry: Principles and Practice. Oxford: Blackwell Scientific Publications. p. 74. ISBN 0-632-02017-2.
20. Snorkel A steam-powered submarine: the Ictíneo. Low-tech Magazine, 24 de ghanta de 2008
21. On Flue gas Condensation by Götaverken Miljö AB
22. "Crystec Technology Trading GmbH, Burn and wet scrubber for process exhaust gas cleaning".
23. "Crystec Technology Trading GmbH, Wet scrubber for process exhaust gas cleaning".
24. "Crystec Technology Trading GmbH, Absorption scrubber for exhaust gas cleaning".
25. Griggs, Mary Beth (20 July 2015). "Mercury Scrubbers On Power Plants Clean Up Other Pollutants, Too". Popular Science. Retrieved 30 July 2015.
26. Lafontaine, Scott; Schrlau, Jill; Butler, Jack; Jia, Yuling; Harper, Barbara; Harris, Stuart; Bramer, Lisa M.; Waters, Katrina M.; Harding, Anna (2015-07-07). "Relative Influence of Trans-Pacific and Regional Atmospheric Transport of PAHs in the Pacific Northwest, U.S.". Environmental Science & Technology. doi:10.1021/acs.est.5b00800. ISSN 0013-936X.
27. How Sulfuric (SO2) waste is chemically removed from emissions at General Chemistry Online
28. Update: Outbreak of legionnaires' disease in Norway traced to air scrubber
29. Coe, H.S.; Clevenger, G.H. (1916). "Methods for determining the capacities of slime-settling tanks". Transactions of the American Institute of Mining and Metallurgical Engineers. 55: 356.
30. "Siltation & Sedimentation". blackwarriorriver.org. Retrieved 2009-11-16.
31. "Siltation killed fish at Batang Rajang - Digest on Malaysian News". malaysiadigest.blogspot.com. Retrieved 2009-11-16.
32. U.D. Kulkarni; et al. "The International Journal of Climate Change: Impacts and Responses » Rate of Siltation in Wular Lake, (Jammu and Kashmir, India) with Special Emphasis on its Climate & Tectonics". The International Journal of Climate Change: Impacts and Responses. Retrieved 2009-11-16.

Index

S

T

U